STRATHCLYDE UNIVERSITY LIBRARY
30125 00079224 1

KU-345-663

ANDERSONIAN LIBRARY
WITHDRAWN
FROM
LIBRARY
STOCK
UNIVERSITY OF STRATHCLYDE

COAL

Its Role in Tomorrow's Technology

A Sourcebook on Global Coal Resources

OTHER PERGAMON TITLES OF INTEREST

ANDRESEN & HAELAND	Hydrides for Energy Storage
BLAIR *et al*	Aspects of Energy Conversion
BOER	Sharing the Sun
HUNT	Fission, Fusion and the Energy Crisis
IAHE	Hydrogen in Metals
KARAM & MORGAN	Environmental Impact of Nuclear Power Plants
McVEIGH	Sun Power
MURRAY	Nuclear Energy
SCHLEGEL & BARNEA	Microbial Energy Conversion
SIMON	Energy Resources
VEZIROGLU	First World Hydrogen Energy Conference Proceedings
VEZIROGLU	Hydrogen Energy System
VEZIROGLU	Remote Sensing Applied to Energy-Related Problems
VEZIROGLU	Energy Conversion – A National Forum
DE WINTER	Sun: Mankind's Future Source of Energy
ZALESKI	Nuclear Energy Maturity

RELATED JOURNALS PUBLISHED BY PERGAMON PRESS

International Journal of Hydrogen Energy

Annals of Nuclear Energy

Progress in Nuclear Energy

Solar Energy

Sun World

Progress in Energy and Combustion Science

Energy Conversion

Energy

Geothermics

COAL

Its Role in Tomorrow's Technology

A Sourcebook on Global Coal Resources

by

CHARLES SIMEONS, M.A.

Industrial Consultant

Former Member of the British Parliament

PERGAMON PRESS

OXFORD · NEW YORK · TORONTO · SYDNEY · PARIS · FRANKFURT

U.K.	Pergamon Press Ltd., Headington Hill Hall, Oxford OX3 0BW, England
U.S.A.	Pergamon Press Inc., Maxwell House, Fairview Park, Elmsford, New York 10523, U.S.A.
CANADA	Pergamon of Canada Ltd., 75 The East Mall, Toronto, Ontario, Canada
AUSTRALIA	Pergamon Press (Aust.) Pty. Ltd., 19a Boundary Street, Rushcutters Bay, N.S.W. 2011, Australia
FRANCE	Pergamon Press SARL, 24 rue des Ecoles, 75240 Paris, Cedex 05, France
FEDERAL REPUBLIC OF GERMANY	Pergamon Press GmbH, 6242 Kronberg-Taunus, Pferdstrasse 1, Federal Republic of Germany

First edition 1978

British Library Cataloguing in Publication Data

Simeons, Charles
Coal.
1. Coal 2. Technological innovations
I. Title
622'.33 TN800 78-40827

ISBN 0-08-022712-0

In order to make this volume available as economically and as rapidly as possible the author's typescript has been reproduced in its original form. This method unfortunately has its typographical limitations but it is hoped that they in no way distract the reader.

Printed in Great Britain by William Clowes & Sons Limited London, Beccles and Colchester

CONTENTS

Appendix

LIST OF FIGURES

Figure No. Page

LIST OF TABLES

FOREWORD

For many years I worked in industry and for much of that time we used coal. Later low priced oil seduced us but that was long before the 1973 oil crisis.

When I entered Parliament in 1970 I was one of the few able to make my contribution as to the needs of industry, and heavy lorries, from personal experience. But of the problems of energy recovery, I knew little.

And yet I was supposed to vote large sums of money to the coal industry without any real knowledge as to its role today, let alone in tomorrow's technology.

However, the impact on my life was fairly immediate since at the 1974 General Election which followed the confrontation with the miners, I departed from the House of Commons no longer an M.P.

A return to Westminster is never easy, but in case the European Parliament should offer opportunities I decided to make good deficiencies in my knowledge with regard to fossil fuels in particular and energy in general. I believed that an additional impetus would be provided were I to write up my findings.

First I examined Oil and Natural Gas Recovery, including their use as a feedstock to the Chemical Industry. I then brought together all the Energy R & D Programmes in Western Europe including those of EEC which indicated that a considerable amount of work was being carried out on coal. Both of these studies were published.

I then took a look at Coal. Most of the books dwell on the evils of the past. A number of learned papers concentrating in depth on a narrow field have however been published, generally as conference proceedings. By seeing for myself and then trying to bring together into one study, the many narrow fields of activity which make up the world of coal, I could be filling a gap in my own understanding of the subject and providing a general review for those interested in coal with statistics and new facts for people working in the industry.

I have visited coal mines, research establishments and suppliers of equipment to the mining industry. I have been in touch with Departments of State responsible for energy and organisations able to provide information on coal from all aspects. I have been most impressed with all that I have seen and I am extremely grateful to all who gave of their time in helping me. I hope my efforts will be of interest.

INTRODUCTION

Coal: its role in tomorrows technology will be that of the innovator commanding a degree of research and development rarely attracted before. It is the only fossil fuel which is likely to be available to the west in abundance over the next few hundred years.

The title might give the impression that this study will examine coal either in its traditional role as a source of heat or else converted to a gas. But this would make the influence of coal upon the technology of tomorrow far too narrow. Coal will continue to present a challenge, in its way, comparable to that of putting men on the moon. In fact the major objective is the opposite: to ensure that men remain on the surface recovering coal, in one of a number of ways, by remote control.

This is the real impact which coal is going to create and of the two it will probably form the greater feat. Governments are always ready to pour money into enterprises which bring national prestige to the nation even though, short term, few people are involved. And yet if the world is to be kept warm and the wheels of industry turning, very large sums of money together with adequate numbers of trained engineers and scientists must be made available for exploration of new reserves, as well as recovery of known deposits.

The processes at all stages need to be examined with these questions in mind.

- What part does coal play in the energy balance of the nation?
- Where are the reserves to be found and what tonnages do they represent?
- Which are the major coal producing nations: what is the state of development of their industries?
- How do the coal producing countries see the needs of tomorrow in terms of their R & D programmes?
- How can technology assist exploration and faster recovery?
- In what form should energy from coal be made available locally and internationally so that best use may be made of existing transport such as the gas distribution networks?
- How is the environment to be improved for those working below ground while that on the surface is preserved from disfigurement.

Mechanisation and remote control are the major problems to be solved.

Recently released figures of World Coal Production show a continuing upward trend of production in 1976.

Preliminary data indicates world anthracite and bituminous coal production in 1976 at 2,468,551,000 tons or 2.3% higher than the 2,412,125,000 tons produced in 1975. The overall world coal output continued a slowly rising trend for most countries except Europe.

This trend can be seen from coal production figures by geographical regions for 1972, 1975 and 1976 as a percentage of total world output expressed in millions of tonnes.

	1976		1975		1972	
	Weight	% World Output	Weight	% World Output	Weight	% World Output
North America	602.9	24.4	591.9	24.5	55.7	25.0
South America	8.6	0.3	8.4	0.3	7.4	0.3
Europe (Excl. USSR)	471.5	19.1	473.0	19.6	466.6	21.0
U.S.S.R.	546.0	22.1	536.0	22.2	499.5	22.6
Asia	684.5	27.7	663.0	27.5	567.0	25.5
Africa	78.9	3.2	70.6	2.9	61.9	2.8
Oceania	76.1	3.1	69.2	2.9	61.8	2.8
Total	2,468.5	100.0	2,412.1	100.0	2,219.9	100.0

In Asia, which has become the largest coal producing region, all the principal countries except Japan showed increased coal production in 1976 over previous years.

However, current world annual consumption of primary energy is of the order of 8,400 million tons of coal - equivalent of which today, coal contributes around 30%. While total world energy consumption could double by the year 2000 unless energetic conservation measures are taken, world production of coal is unlikely to exceed 4000 million tons, the equivalent of an annual growth rate of some 2% per annum only.

To achieve this tonnage much research and development world-wide will be needed particularly into improved methods of extraction which will leave less roof coal behind. Seams which previously could not be worked, must now be mined, while ways of burning high sulphur coals must be found without infringing the stringent anti-pollution laws.

Seams from 50 cm to several metres thick are involved both close to the surface and at depths in excess of 1200 metres.

Deposits which are undisturbed or faulted; or with seams ranging from the horizontal to steep inclines - even the vertical must be mastered.

Gas emissions, heating of the working atmosphere by hot exposed rocks and from highly rated equipment installed at the surface, need to be tackled.

Short term the objective must be to remove men from the coal face. Long term it involves everyone working in coal including preparation, conversion, transport and ultimate use, particularly in the Chemical Industry.

The challenge is there. In fulfilling its role, coal will have an immense influence on the course of tomorrow's technology. It will influence every field of scientific and technological endeavour from engineering to explosives, science to safety and medicine to methodology.

It is a challenge we ignore at our peril. It is one which we must win.

Chapter 1

ENERGY DEMAND

Three vast national groupings, U.S.A., Japan and Western Europe account for nearly two thirds of world energy demand. Oil naturally figures strongly in total needs. Back in 1973, there was a gap between production and self sufficiency of some 1600 mtce, but as the pressure created by these three blocks for world resources of gas and oil increased, prices rose and consumption fell.

In 1974, world consumption of energy according to figures presented to the U.K. National Energy Conference totalled 9240 mtce (5600 mtoe). However, the United Nations Organization statistics for 1976 showed this figure to have fallen, in 1975, to 8020 mtce. Industrial stagnation and attempts at conservation had played their part. Now, four years on, the industrialized world runs predominantly on oil. Natural gas and coal come next with water power and nuclear energy supplying only a small part of total demand. Clearly fossil fuels - coal, gas and oil - are vitally important. There is in fact no shortage of supply world wide.

As will be shown in Chapter 2, proven reserves of oil and gas recoverable with current technology are sufficient to meet demand until 1990. There are also prospects of further discoveries which will take us well beyond that point. Coal is even more plentiful with reserves for at least a further century.

Unfortunately these oil and gas reserves are not evenly distributed and are not easy to obtain. Those countries using oil predominantly do not possess sufficient, while those owning the oil at present often do not use it.

It is in part a political problem which governments have to solve. As the Workshop on Alternative Energy Strategies held in 1977 concluded:

"The interdependence of nations in the energy field requires an unprecedented degree of international collaboration in the future. In addition it requires the will to mobilize finance, labour, research and ingenuity with a common purpose never before attained in times of peace."

And yet, a Gallup survey taken again in mid 1977 in the United States showed that a very large number of Americans did not even know that the United States imported oil - running then at a level of some sixty per cent. It is clear that the consumption of energy runs parallel with the level of the Gross Domestic Product so far as developed countries are concerned.

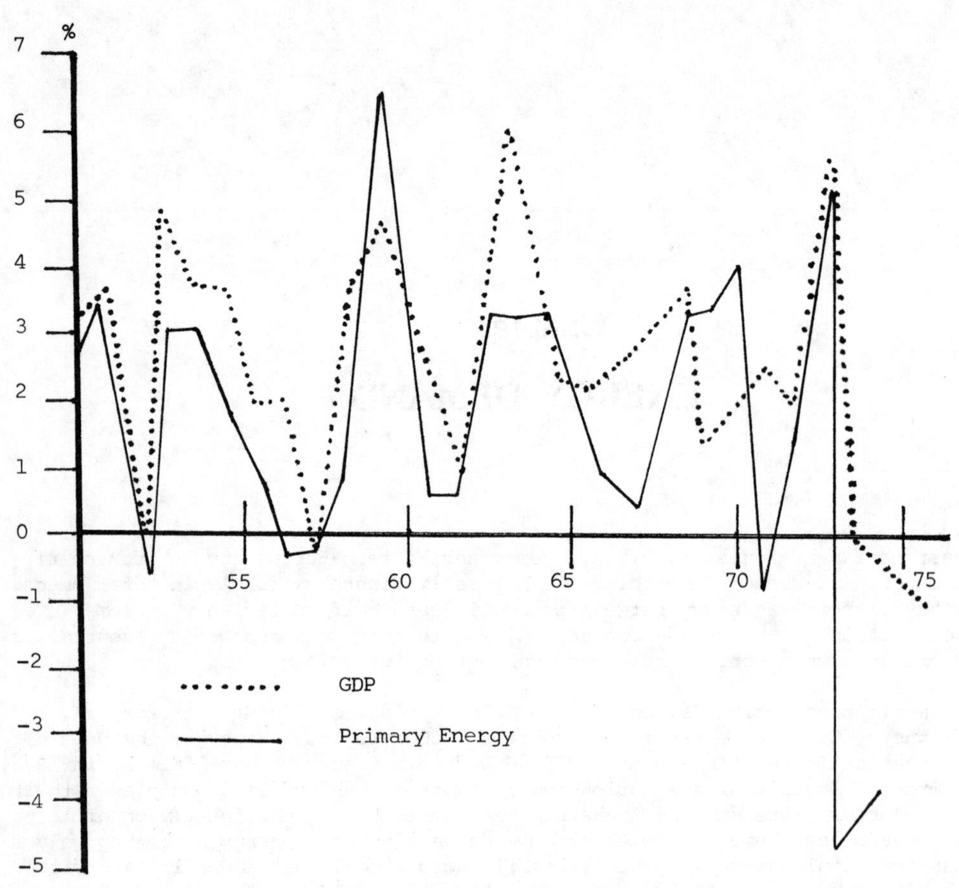

Fig. 1. Growth of GDP and primary energy consumption Britain 1950-1975.

Figure 1 illustrates this clearly so far as Britain is concerned, where production remained static between 1974 and 1978 - although this forms part only of the GDP.

Recently the Cavendish Laboratory, Cambridge England, made projections of energy demand growth rates for world regions employing assumptions for economic growth as the basis, using high and low levels. These are shown in Table 1.

The following regions, sectors and fuel or energy carriers were surveyed.

World Region	Economic Sector	Fuel or Energy Carrier
N. America	Transport	Coal
W. Europe	Industry	Oil
Japan	Residential	Gas
Rest of World (non communist)	Non energy use of fuel	Electricity

Although one energy source and one sector of activity only were examined, it would have been quite practicable to have extended the study.

The next stage is to examine the Community forecasts.

TABLE 1 Projected Energy Demand Growth Rates for World Regions

Region	Energy AAPG 1960-72	Energy AAPG 1972-85 Unconstrained		Energy AAPG 1985-2000 Unconstrained	
		High	Low	High	Low
N. America	4.1	2.6	1.9	2.6	1.9
W. Europe	5.2	3.3	2.6	2.9	2.2
Japan	11.2	5.2	3.6	4.1	2.8
Rest WOCA	6.8	6.3	5.0	5.0	3.8
WOCA	5.2	3.6	2.8	4.0	2.5

In table 1 WOCA indicates world outside communist area. These projections for potential energy supply to 1985 take into account the expected surplus capacity for oil production during this period which is expected to inhibit the growth of alternatives. From 1985 to 2000 a fast expansion is assumed for both coal and nuclear, although rates of expansion are constrained by the lead times for developing the industries. Although their projections were extended beyond 2000, to become involved in such crystal gazing might be tempting fate and, in any case, goes rather beyond "Tomorrow's Technology" in respect of coal.

Examination of the European scene indicates the great future for coal as illustrated in tables 2 and 3 with respect to the Community.

TABLE 2 Community Energy Situation 1976 (mtoe)

	Production	Net Imports	Consumption	%
Solid fuels	184	23	207	22
Oil	22	520	542	58
Natural gas	144	12	156	16
Hydro and geothermal	25	1	26	2
Nuclear	21	-	21	2
Total	396	556	952	100
%	42	58	100	

It is clear from table 2 that 74% of present energy consumption within the Community comes from sources with a limited life span, at the best up to 2000 A.D. The problem will begin before then, when demand begins to exceed the rate of recovery. The latter will become more difficult as wells reach exhaustion, pressures fall and exploration moves to deeper waters. Forecasts for 1985 within the Community are listed in table 3.

The great question to be answered is the role which Nuclear Energy is to play. Table 3 indicates a considerable increase from 2% of the total in 1976 to 11% in 1985. Even so, how great a stumbling block will political opposition to new nuclear plants prove to be? Will proposed fuel processing plants suffer a similar fate? Fortunately the British Windscale enquiry came down in favour of a processing plant being built there which both the British Government and House of Commons later endorsed so that the scheme can go ahead.

TABLE 3 Community Energy Forecasts for 1985 (mtoe).

	Production	Net Imports	Consumption	%
Solid fuels	184	36	220	17
Oil	110-160	555-490	665-650	52-51
Natural gas	143-158	79	221-237	17-18
Hydro & geothermal	31	4	35	3
Nuclear	140	-	140	11
Total	608-673	674-609	1282	100
%	47.4-52.5	52.5-47.5	100	

It is surprising to note that coal production is not expected to rise within the Community, imports still being regarded as necessary. Dependence upon imports will increase in total tonnage although the proportion will fall from 58% to around 50%. Originally it was intended that the dependence level should be down to 40% - clearly an impossible target.

Progress since 1958 is shown in table 4.

TABLE 4 Energy Dependence - European Community

1958	The six only	29%
1963	"	43%
1968	"	57%
1973	The nine	61%
1974	"	61%
1975	"	57%
1976	"	58%
1985	Forecast - mean	50%

It is clear that the nuclear programme will not be sufficiently advanced to assist materially by 1985 leaving a very considerable dependence upon coal to improve the "imports situation".

A close examination of the needs of individual members of the nine produces the pattern shown in Table 5 with regard to imports of fuels as a percentage of total energy needs for 1974 and forecasts for 1985.

It should however not be considered in isolation from the way in which the oil and gas element will be replaced.

Denmark, Italy, Luxembourg and the Netherlands produce little or no coal. They will have to make good their oil and gas deficiencies from renewable sources of energy - unlikely to provide more than a small percentage - or coal obtained from countries which enjoy abundant reserves. These include the United States, U.S.S.R., China and India. The United States has troubles of her own while lines of communication from the others are long and not totally divorced from political risk, although this may be less than the potential security problems of the North Sea.

It is worth noting that by 1985 Germany will be obtaining gas from Netherlands, Norway, Iran and U.S.S.R.

TABLE 5 Total Imports of all Fuels as a Percentage of Total Energy Needs for 1974 and 1985

	1974	1985
Belgium	89.0	78.2
Britain	50.8	17.2 - nil
Denmark	99.6	98.1
France	79.0	63.2
Germany	52.9	56.0
Ireland	82.3	76.5
Italy	82.8	69.5 - 71.9
Luxembourg	99.4	83.5
Netherlands	6.0	34.5

The figures listed in Table 5 for 1985 indicate a reduced dependence upon imports except for Germany and the Netherlands. The latter will require greatly increased oil imports, although some 33.6% of energy requirements will be exported as gas - a slight increase over 1974. Britain could well be self sufficient or even exporting oil, while Germany will be importing 30% more oil.

The percentage of solid fuel imports relative to total solid fuel needs are shown in Table 6.

TABLE 6 Solid Fuel Imports as a Percentage of Total Solid Fuel Needs

	1974	1985	Total Tonnage millions
Belgium	57.7	54.5	17.3
Britain	< .01	nil	141.6
Denmark	100.0	100.0	4.9
France	47.1	63.3	47.2
Germany	nil	nil	138.2
Ireland	35.3	31.5	3.2
Italy	96.7	97.7	20.5
Luxembourg	100.0	100.0	3.3
Netherlands	82.8	100.0	11.3

From Table 6 it is clear that France will be importing considerably increased quantities of coal, while by 1985 Britain and Germany should be in a position to export. The Netherlands, where coal-mining ceased in 1976, will also require increased imports of solid fuel by 1985.

The dependence upon coal within the Community will have fallen from 24% in 1974 to 22% in 1976. But it has been agreed that this figure shall not be allowed to fall below 17% as indicated in Table 3. This will amount to an increase of 19 million tonnes in terms of coal used.

The overall reduction in energy consumption in 1976 as compared with 1973, affected both oil and coal whereas natural gas appreciably increased its contribution with North Sea oil on stream.

UNITED STATES

U.S. demand for energy is increasing while the available domestic supply of oil and natural gas has been declining and will continue to do so. To meet increasing demand the United States has turned more and more to imports which has resulted in increased vulnerability to interruption in supply.

This has led to a need to develop natural gas policy in those terms:

- a need to bring the supply of natural gas into balance with demand
- a reduction in the use of natural gas without a complementary increase in oil imports

The coal policies proposed by the Administration have been designed specifically to achieve these ends through:

- natural gas pricing provisions
- new oil pricing policy
- oil and gas consumption taxes

To achieve these ends the plan has four major features

1. Conservation and increased fuel efficiency.
2. Rational pricing and production policies.
3. Substitution of energy sources in plentiful supply for those experiencing shortage.
4. Development of non-conventional technologies for the future.

Coal Programme

The coal programme will save the equivalent of about 3.3 million barrels of oil a day by 1985. This programme takes into consideration new electricity utilities as well as industry both in respect of coal and other fuels.

The administration has basically placed a total ban on new oil and gasfired electric facilities although there will be exceptions where specific economic reasons or ability to maintain a service precide. Environmental problems may also be a factor. Utilities and industrial facilities will be asked to convert to coal without impairing air quality standards. In areas already experiencing air pollution problems, to a significant extent, it may be necessary to continue burning oil to safeguard health.

This policy systematically enforced, may well lead to investment in new energy facilities. All new facilities including those which burn low sulphur coal will be required to employ the best available control technology. This cannot take place overnight and therefore a tax will be levied on oil and gas consumption beginning in 1979 for industry and 1985 for utilities.

The tax would be imposed only on large users of oil and gas amounting to 2000 firms out of a total of 100,000. These 2000 firms are believed to consume about 90% of all industrial oil and gas in the U.S. The idea is that the money needed to pay the taxes would be better used in coal conversion.

The programme is designed to encourage the use of new technologies such as low and high Btu coal gasification, fluidised bed combustion and other technologies designed to burn coal in an environmentally and economically practical way.

The plan is expected to restore some stability to the coal industry in the knowledge that there must be a growing reliance upon coal.

When the Energy Programme has been accepted by Congress, it will be easier to forecast needs ahead. The energy balances for a number of countries including the United States are shown in Appendix A.

Chapter 2

COAL — SOURCES AND RESOURCES

Before discussing reserves of coal and later the particular problems of their recovery, it may be helpful to examine the way in which coal is formed and in turn the problems encountered in mining it.

Coal is made up of plant material. This is borne out by the fact that imprints of leaves and stems of plants are often found in the roof of a coal seam. Also, in the floor or rocks just below the seam, dark markings may be seen, the effect of former tree roots. Chemical analysis too indicates coal to be in origin a type of wood.

This change has not been sudden. It has taken some tens of millions of years to occur. Trees grew in the swamps, but when dead formed a layer of decaying vegetable material - leaves, branches and trunks - often many feet thick. This land then sank below the water with sand and mud carried down river being deposited on top until the water became sufficiently shallow for trees to grow again and form another layer of peat. Similar forest swamps are also found in the deltas of tropical rivers today. Many of the lands forming countries as we know them now did not take on the shape commonly shown on modern maps. Many contained vast shallow estuaries bordered by hilly land from which great rivers flowed into the estuaries. The present world coal fields originally formed areas that were sinking slowly at a rate which enabled sand, mud and clay to be deposited by the rivers on the bottom of the estuaries which remained at a fairly consistent depth of water. From time to time sand and mud nearly filled up a shallow estuary turning it into a swamp where trees and other vegetation could grow. This state lasted for many centuries producing a type of peat.

The process was repeated until after millions of years it came to an end. Areas of land and sea then changed, often with thousands of metres of rocks being formed on top of the layers of peat, sand and clay. This build up resulted in enormous pressures accompanied by an increase in temperature due to the depth involved. Peat layers slowly became converted into coal. The other layers hardened too, sand being converted to sandstone and clay into shale.

But seams of coal will vary in depth, thickness and the angle at which they run. It is this variation which creates the problems encountered in mining.

In many instances, as already indicated, material from higher land masses has been redeposited in lower areas, causing distortion into folds to produce synclines or anticlines as shown in Fig. 2.

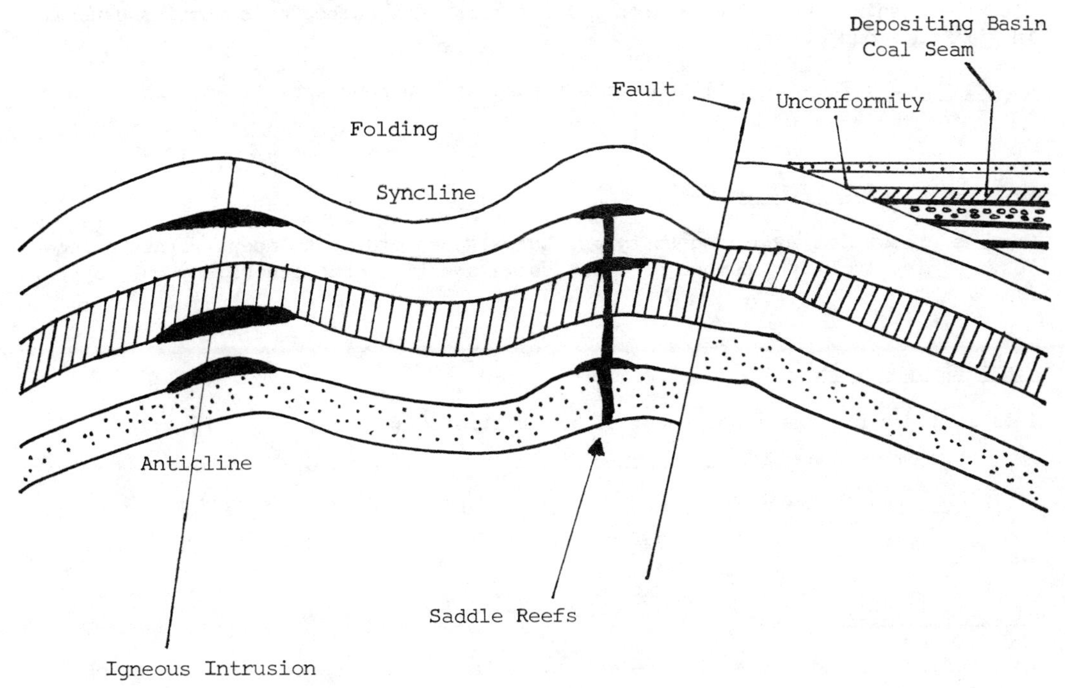

Fig. 2. A typical folding process.

Where the folding is severe the rock becomes heavily fractured, particularly where it is under tension on the crest of an anticlinal stratum or the base of a synclinal stratum. The weakened zones provide the path for intrusive rocks and mineral rich solutions to permeate into the host rock producing ore bodies. Many ore bodies have a typical form, i.e. a saddle.

TYPES OF COAL

Peat. This occurs from the first stage of coal formation. It is to be found in many parts of the world particularly Ireland. When dug for a fuel, it consists of partly decayed reeds and mosses growing in bogs.

Brown Coal. This results from the first stage in the process of change in buried peat. It is brown, crumbles and may appear to be composed of decayed woody-like material. Little is found in Britain, but large quantities occur in Australia, South Africa, Germany, Poland, U.S.S.R. and the United States. This type of fuel is sometimes called lignite, although some lignites are black.

Bituminous Coals. These are the commonest for use in houses and factories. They are always black, and composed of bands or layers which vary in appearance from bright and ceramic-like to matt and sooty. It generally fractures easily into convenient blocks in set planes which assists in the winning of coal in rather the same way that there is grain in wood. This plane is often called a cleat. It occurs in many countries.

Anthracite. This is a type of coal which has progressed further along the process of change from peat. It is very hard, has lost the banding, breaks into small blocks and shines brightly. Anthracite occurs in a number of countries generally in limited amounts.

Cannel Coal. This is dull and hard-burning with a long smoky flame like a candle. It is without bands or cracks.

PROPERTIES OF COAL

Coal is classified according to types, but also according to needs within categories. This involves a number of tests, some specific, but others are only approximate.

Proximate Analysis

This includes precise measurement of the following :

percentage moisture | volatile matter
ash | fixed carbon
calorific value | sulphur content

Ultimate Analysis

This series of tests is much more extensive involving percentage by weight of elements - usually:

carbon | hydrogen
oxygen | nitrogen
sulphur

Moisture Content

This investigation involves two tests under set conditions:

(a) Free percentage of moisture lost when naturally moist finely ground coal is allowed to reach equilibrium with the atmosphere at 15.5°C.

(b) Fixed percentage of moisture present in air dried coal.

Ash

A known weight of coal is taken and heated in a furnace in air at 800°C.

Volatiles

This test involves the percentage of products given off when coal is heated in a covered crucible to a temperature of 925°C under standard conditions.

Fixed Carbon

This figure is arrived at by deducting the sum of the percentages of ash, volatiles and moisture from 100.

Calorific Value

The gross calorific value of coal as determined in a bomb calorimeter using the following basis for reporting:

1. as received
2. dry
3. moisture and ashfree (daf - dry ash free)
4. mineral matter free (mmf)
5. moisture and mineral matter free (dmmf - dry mineral matter free)
6. calorific value (cv)

DISTURBANCES IN COAL SEAMS

It is a fact that the rocks forming the outer layers of the Earth's crust - some 30 miles in depth - are always under stress and strain drawing some parts together, forcing others apart, raising areas higher relative to the surroundings and causing others to fall. These movements are immensely slow, but over a period of millions of years, they make a considerable impact. Where these movements are perceptible earthquakes may result. Volcanoes also result from earth movements. Any change is of importance to the miner: it may take one of a number of forms.

The simplest form of movement is tilting up to the vertical in some instances. Where the tilt is downwards it is called the "dip", the upward slope often being termed the rise. The direction at right angles to the dip is called the strike.

Strata may not only be tilted: they may also be bent or folded taking on the form of a wave in which the coal seam will form one or more layers. That part forming the crest of the wave is called the anticline, while the trough is known as the syncline.

However, instead of bending, the strata may break. This could result from having bent too far. Such breaks are called "faults", as shown in fig. 3.

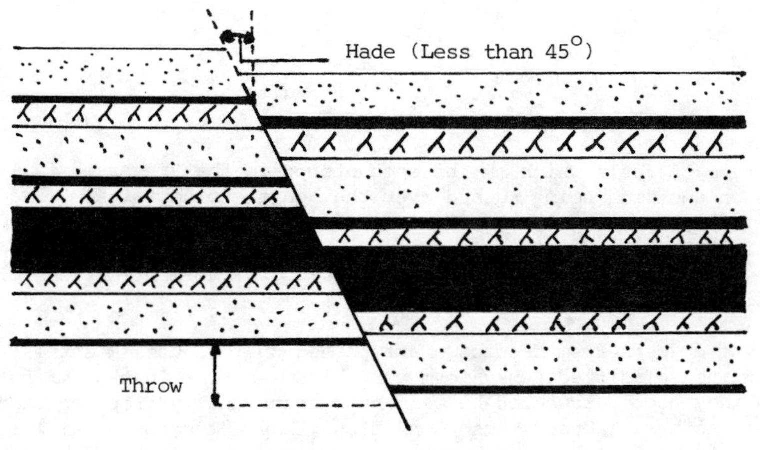

Fig. 3. Normal fault.

The miner discovers that the seam which he is working suddenly stops due to a "fault", but is continued on the other side at a higher or lower level. The displacement is known as a "throw", amounting to a few inches or possibly as much as tens of metres. Where the seam continues at a higher level, it is known as an "upthrow fault", while at a lower level it is called a "downthrow fault". Rocks in the vicinity of the fault are frequently crusted and broken while the surfaces are often highly polished because one side has slipped under enormous pressure over the other. Such polished surfaces are known as "slickensides". The angle between the fault plane and the vertical is known as the 'hade' which in the case of the normal fault is less than 45°. It is nearer to the vertical than horizontal. The slope is always towards the lower or downthrow side.

Very often there may be more than one fault resulting in a number of parallel faults all "throwing" in the same direction. This is known as step faulting as shown in fig. 4.

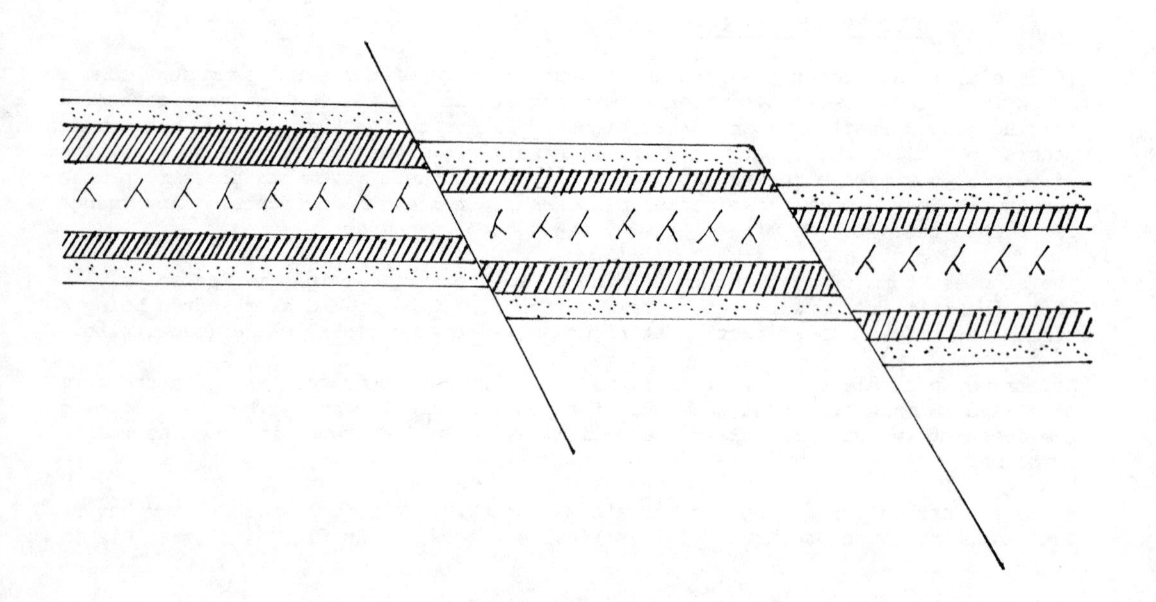

Fig. 4. Step faults.

A further type of fault which may be encountered is the "reversed fault" caused by the strata on one side being pushed over that on the other side. Figure 5 shows this effect clearly.

Although a slight deviation from faults it is useful to mention some of the other abnormalities which can occur.

Sometimes hot objects from deep down are pushed up into Coal Measures. After they have cooled and solidified, they form very hard walls. Instead of cutting across the strata they become squeezed between beds occasionally dropping suddenly from one lead to another. Under these conditions they are known as sills. Other forms of abnormal strata include a wash-out, or when the seam is missing with the rock forming a ridge it is called a horseback.

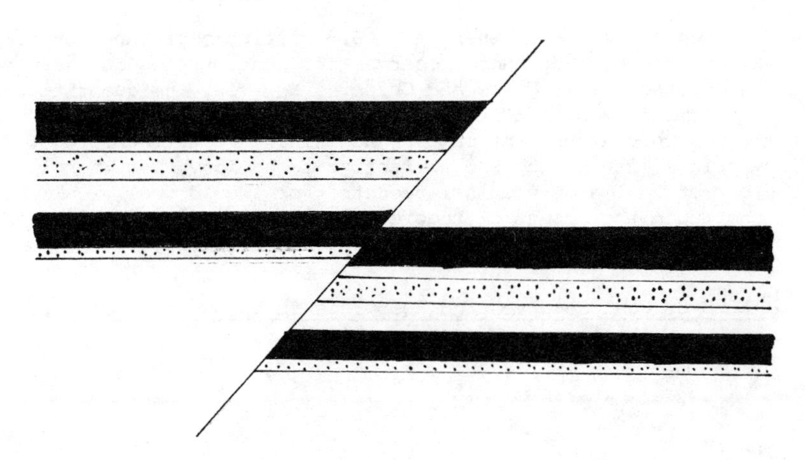

Fig. 5. Reversed fault.

A "trough fault" occurs when two breaks in the strata allow the rocks between them to drop, forming a trough.

Other variations in the strata make complications for the miner when the seam thins out suddenly, only to recover just as quickly. Alternatively, molten rocks may interrupt the seam in a number of ways.

Resources

Every country with solid fuel resources needs to know where the resources are to be found, the type of coal and its characteristics. The quantities available also need to be known, including the extent to which they are accessible.

Table 7 lists resources published following a survey by the World Energy Conference 1974, the most thorough study of its type in recent years.

TABLE 7 Resources as at 1974 by Continents and Nations with Major Resources (Millions of Tonnes).

Country or Continent	Reserves		Total Resources
	Recoverable	Total	
USSR	136,600	273,200	5,713,600
China, P.R. of	80,000	300,000	1,000,000
Rest of Asia	17,549	40,479	108,053
United States	181,781	363,562	2,924,503
Canada	5,537	9,034	108,777
Latin America	2,803	9,201	32,928
Europe	126,775	319,807	607,521
Africa	15,628	30,291	58,844
Oceania	24,518	74,699	199,654
World Total	591,191	1,402,273	10,753,880

From Table 7 it can be seen that there are 10.8 million megatonnes of solid fuel reserves. Ninety per cent of these reserves are concentrated in three countries; the United States, the Soviet Union and China. However, the quantities likely to be recovered economically are limited to 1.4 million megatonnes of which only 0.6 megatonnes are feasible at present prices and using current technology. However, rising energy prices, new mining and extractive technology, together with developments in the use of coal as a chemical feedstock or liquid fuel, would have a considerable impact upon the amount of recoverable reserves actually recovered. Table 8 shows the position with regard to resources in Europe.

TABLE 8 Coal Reserves and Resources in Europe as at 1974. (Millions of Tonnes)

Nation	Reserves		Total
	Recoverable	Total	Resources
Western Europe			
West Germany	39,571	99,520	286,150
Netherlands	1,840	3,705	3,705
France	458	1,407	1,407
Belgium	127	253	253
Austria	64	148	177
Total	42,060	105,033	291,692
Southern Europe			
Yugoslavia	16,870	17,976	21,751
Spain	1,643	2,202	3,562
Greece	680	908	1,575
Italy	33	110	110
Portugal	33	42	42
Total	19,259	21,238	27,040
Northern Europe			
United Kingdom	3,871*	98,877	162,814
Finland (Peat)	4,290	33,000	33,000
Sweden (Peat)	30	9,460	9,490
Iceland (Peat)		2,000	2,000
Denmark (Peat)	20	561	581
Ireland (Peat)	418	422	448
Norway	2	2	152
Total	8,631	144,322	208,485

* For the 1977 World Energy Conference revised to 45,000 million tonnes.

(Table 8 continued)

TABLE 8 (Continued)

Nation	Reserves Recoverable	Reserves Total	Total Resources
Czechoslovakia	6,363	13,774	21,430
Hungary	1,675	3,350	6,400
Bulgaria	4,387	4,387	5,230
Romania	1,150	3,970	1,960
Total	61,515	94,555	125,670
Total Europe	131,465	365,148	652,887
Eastern Europe			
Poland	22,640	38,874	60,600
East Germany	25,300	30,200	30,050

Most of the European countries excluding Scandinavia possess coal resources. The Scandinavian countries own larger quantities of peat, valued much more highly as oil prices rise and uncertainties with regard to other fossil fuels increase.

In Western Europe 98 per cent of coal resources are found in West Germany. While France, Netherlands and others have run down their mining industry and boast of but small reserves. Even so those reserves which do exist could be significant and make a useful contribution to the general European need.

In Britain, only 4 per cent of reserves are considered to be recoverable, but strenuous work on an increased exploration programme is expected to discover new underground reserves.

Turning to Europe, the Community has been forced to re-examine plans for coal in order to reduce dependence upon oil, much of which is imported - other than in the case of U.K. and Norway.

Countries not specifically mentioned, but worthy of comment include:

Australia. 74,341 million tonnes, out of a total of all types of reserves amounting to 198,567 million tonnes are known to exist. Of those, 13,770 tonnes are recoverable hard coal with 10,555 million tonnes of poorer types.

India. There are 23,139 million tonnes from a total of all types of reserves of 82,937 million tonnes of which 10,643 million tonnes represent recoverable hard coal, 897 million tonnes only coming from lower ranked coals.

South Africa. Reserves amount to 24,224 million tonnes of all types recoverable out of a total of 44,339 million tonnes. Of this, 10,584 million tonnes represent hard coal.

Hard Coal Production

World production compiled on a similar basis appears in Table 9 in respect of hard coal. It can be seen that five of those listed consist of groupings of countries or areas which between them account for 92.5% of the total hard coal production world wide.

TABLE 9 Hard Coal Production by Continents and Nations - 1976
(Millions of Tonnes)

Country or Continent	Hard Coal Production	Percentage of World Production
USSR	546.0	22.2
China	480.0	19.5
Rest of Asia	194.3	7.9
United States	580.0	23.6
Canada	20.8	0.8
Latin America	8.7	0.5
Europe	470.7	19.3
Africa	78.9	3.1
Oceania	78.9	3.1
	2458.3	100.0

The countries of importance listed with respect to resources, but not mentioned specifically in the tables, produced the following quantities of hard coal during 1976.

Australia. Production, included in the figures for Oceania totalled 85 million tonnes of coal.

India. 103 million tonnes of coal which were equivalent to 4.2% of world production was mined during 1976.

South Africa. The total tonnage mined during the year amounted to 74 million tonnes or 3.0% of the world total.

Looking ahead to 1985 the World Energy Conference held in Istanbul anticipated that coal production would rise from the 1975 level of 2384 million tonnes to 3596 million tonnes by 1985 reaching 4000 million tonnes by 2000 AD. Coal now accounts world wide for 32% of all energy needs. To maintain this position, it will be necessary to increase production by 3.6% annually. A 40% share would require an increase of 4.8% per annum. This is unlikely in most countries bearing in mind the long lead-in time for new mines to be opened, particularly the building up of the infra structure including transport and people to work there.

Assessments of national production levels for 1985 were forecast in this order - in millions of tonnes

China	920
USSR	851
United States	842

China will overtake the USSR and the United States, while the latter will slip back to third place.

A factor in the USSR's favour is the developed use of hydraulic methods of mining which will be referred to in the next chapter.

Table 10 illustrates the production levels of hard coal within European countries during 1976.

TABLE 10 Hard Coal Production in Europe - 1976 (Millions of Tonnes)

Nation	Hard Coal Production	Percentage of World Production
Western Europe		
West Germany	89.3	3.6
France	21.9	0.9
Belgium	7.2	0.3
	118.4	4.8
Southern Europe		
Yugoslavia	0.6	<0.1
Spain	11.1	0.4
	11.7	0.4+
Northern Europe		
United Kingdom	122.2	5.0
Czechoslovakia	28.3	1.1
Hungary	2.9	0.1
Bulgaria	0.2	0.1
Romania	7.1	0.3
	38.5	1.5+
Poland	179.3	7.3

Table 10 is much shorter containing far less countries than those listed in Table 8. This is because very few of the smaller countries with limited coal reserves and resources have a coal industry. They tend to be importers.

Table 11 illustrates this movement of coal, so far as exports are concerned. It is interesting in that it does not feature China, one of the large producers, as an exporter. On the other hand, the United States lead by a considerable amount, with Poland second and Australia, not a large producer, third, disposing of just under 50% of production. The Soviet Union comes next.

There are also considerable differences between 1960 and 1975.conditions. In 1960 the United States was the major exporter too but Germany featured second just above Poland. While German coal exports have fallen over the period, those from Poland have soared. In fact, they more than doubled.

Canada is a particularly high flyer having risen from the rate of lowest export performance in 1960.

TABLE 11 World Coal Exporters 1960 and 1975
(Thousands of Tonnes)

Exporters	1960	1975
United States	34,456	60,238
Canada	774	11,695
Germany	17,974	14,709
France	1,419	553
Belgium and Luxembourg	2,238	357
Netherlands	2,165	237
United Kingdom	5,547	2,182
Soviet Union	12,300	26,143
Poland	17,497	38,348
Czechoslovakia	2,195	3,666
Australia	1,584	32,422
South Africa	950	2,687
Others	3,664	1,381
Total Exports	102,763	194,618

The receiving countries figure in Table 12. They are much more numerous featuring a number of exporters in the dual role.

TABLE 12 World Coal Imports 1960 and 1975
(Thousands of Tonnes)

Importers	1960	1975
Canada	12,299	15,256
Brazil	928	2,830
Chile	351	164
Sub Total	13,578	18,250
Belgium and Luxembourg	4,148	6,752
France	10,112	17,293
Germany	6,705	6,244
Italy	9,739	12,852
Netherlands	6,868	4,104
United Kingdom	-	5,083
Ireland	1,676	690
Denmark	3,954	4,132
Sub Total	43,202	57,150

(Table 12 continued)

TABLE 12 Continued

Importers	1960	1975
Austria	3,720	2,583
Finland	2,937	3,845
Greece	209	781
Norway	315	456
Portugal	350	434
Spain	307	3,974
Sweden	1,929	1,632
Switzerland	1,952	123
Yugoslavia	1,383	2,310
Sub Total	13,102	16,138
Bulgaria	-	6,379
Czechoslovakia	2,402	5,182
German Democratic Republic	8,028	6,440
Hungary	1,431	1,438
Poland	796	1,096
Romania	416	2,527
Soviet Union	4,776	9,818
Sub Total	17,849	32,880
Japan	8,292	62,107
Others	6,740	8,093
Total Imports	102,763	194,618

It can be seen from Table 12 that Japan at 62 million tonnes is the leading importer by a very considerable amount. France comes next at 17.3 million tonnes, then Canada at 15.3 million tonnes - some 4 million tonnes more than are exported. Italy is a large importer at 12.8 million tonnes, the Soviet Union with 9.8 million tonnes only one third of the quantity exported.

Belgium and Luxembourg, Bulgaria, the Federal Republic of Germany and the Democratic Republic all exceed 6 million tonnes. Surprisingly, the United Kingdom is not far behind at 5 million tonnes - perhaps a reflection of industrial unrest in a coal rich country.

Finally, the whole series of constituents are brought together in one table indicating total world recoverable reserves, consumption figures for 1975 and the duration of reserves at varying growth rates. Those appear in Table 13, from which it can be seen that world consumption of coal and lignite is rather lower according to UNO statistics than the production figures shown in Table 9.

This chapter has set out to show the way in which coal is formed, and the types of coal which may result from different formations, together with the characteristics which are important to the user. The uneven pattern of the incidence of coal seams and the way in which continuity or even access are limited by changes in geological formations have been included here, so that the problems of recovery may be highlighted.

Lastly, the balance between production and trade has been reviewed so that the world balance contained in Table 13 might bring all these factors, including resources, together.

TABLE 13 Duration of Recoverable World Energy Reserves
(Thousand million tons of coal or coal equivalent)

		Total recoverable reserves (including probable and undiscovered)	1975 consumption	Duration of reserves in years at exponential consumption growth rates of: 0%	2%	4%
Oil		366	5.64	82	49	37
Gas		269		155	72	51
Coal and Lignite at recovery rate of	10%	1015	2.36	379	108	71
	50%	5073		1895	185	110
Uranium up to 30 /lb	in thermal reactors	93				
	in fast reactors	4612 (excluding CPEs)				
			8.00			

Consumption UNO Statistics 1976

Reserves etc. U.K. Dept. Energy

With the possibility that world reserves of coal will last for well over 1000 years - if the rate of recovery can be raised to 50% - there is clearly a bright future for the mining industry and those who supply it with the latest technology and means of extraction and recovery.

Chapter 3

THE STATE OF THE INDUSTRY

Before examining the equipment available today and that required for tomorrow, it is necessary to review present practices in a number of mining nations, including their research and development programmes.

This is best done with reference to specific mines, where possible, to illustrate methods being used today, together with the needs for the years ahead, according to the particular problems and conditions in the national mines. These will vary from country to country and from one mine to another. They will include problems of thin seams, faults, and other defects in the strata mentioned in Chapter 2, plus gas and dust. Also coals with a high rock or shale content, all of which will affect those working on the coal face and below ground, as well as the degree of preparation required before the coal can be delivered to the consumer.

This chapter now examines the main sectors of coal mining,roadways, recovery by both conventional as well as hydraulic methods, conveying of men and coal preparation, and health and safety. The review begins with U.S.S.R. which claims to lead the world in the number of mechanical coal faces and levels of output.

USSR

Production in millions of tons (mts) - achieved and forecast

1976 - 712

1980 - 805

Number of mines:

Total - 543

Large mines - 72

Large mines include those with an annual production of 131,000 tons and over.

While it is difficult to balance all the statistics, it may be of interest to note that according to reports production included 542 mts hard coal, 163 mts lignite, 475 deep-mined and 229 mts surface-mined. Growth of open cast represented 30.9% in 1974, rose to 32.2% in 1975 and 32.6% in 1976 for eastern areas of USSR. There are 61 open pits overall. Preparation plants under the control of the Ministry of Coal Industry processed 284.5 mts in 1976 yielding 162.6 mts clean coal.

Proved and possible reserves of coal, 70% of which are concentrated in the eastern areas, are said to be the equivalent of 420,400 mts.

Mining Methods

Long wall mining predominates, minimum losses being achieved through the employment of narrow web techniques coupled with widespread use of power supports. This method of recovery, together with the use of mechanical loading, has increased from 19% to 84.8%. Mechanized complexes represent an increase from 8.3% to 63.7%, while greater use of heading and tunnelling machines, particularly where harder rocks are concerned, has brought about an increase in the share of total lengths of all roadway driven in this way from 6.8% up to 32%. Motor capacities and the performance resulting from this increase in available power of the main mining machines has increased two to three times while the quality and reliability of the equipment is claimed to be better. 660 volt systems have been introduced while a transfer to 1140 watts is in progress. There are 1000 mechanised complexes in the Soviet Union with an average daily quota for faced equipment of 860 tons run of the mine coal, half of these complexes reaching or exceeding 1000 tons output per day. The best seams average 3000 to 5000 tons, while some are said to achieve even 8000 tons.

The system is divided between long pillar methods and room and pillar methods using short faces. Initially, three inclines are driven into the seam which are followed closely with air supplies and gate roads.

The area to be worked is divided into blocks 340-450 metres long and 150-250 metres wide. Every 6 metres, a road is driven across the block, the contained area being extracted by shearers or using hydraulic monitors.

Raspadskaya colliery. This has vertical shafts sunk to 320 metres with seams which dip in from the face to the bottom of the shaft. As development proceeds, the headings are driven out 1½ km from the shaft into the seam by PK3, 4PU and GPK heading machines, after which long wall faces between 100 and 150 metres in length are headed out and worked retreat fashion back to the pit bottom.

The coal is won by KSH3M and KSHIG shearers loading on to double chain armoured face conveyors. Roof support is provided by MK30P, M87, 1MKM, 2MKE, KM130 and the new 2M120 self-advancing powered supports. These will vary as to configuration, but are adaptable to the varying conditions of seam thickness, hardness of roof and softness of the floor.

Transportation of Coal

This has been updated with a reversion to advanced flow line schemes incorporating conveyors with standard belts up to two kilometres long and capable of carrying 1200 tons per hour.

Raspadskaya colliery. It deals with coal in this way. The coal is put on to armoured faced conveyors with a capacity of 1000 tons per hour and then on to trunk belts capable of dealing with over 1300 tons per hour. These deliver the coal into a storage bunker in the pit bottom, via skips, and then up the shaft to 40,000 ton storage bunkers, on the surface, from which it is loaded into rail cars. A second loading point rated at 3000 tons per hour has been completed and now the aim is a still larger complex with a capacity of 6000 tons per hour.

There is a separate shaft for air, men and materials. The men are carried in ski-lifts over one kilometre in length, while materials are hauled by bogies on rope.

Open Cast

Some 40% of the overburden is removed by drag lines and buckets with a capacity of between 10 and 15 m^3 - sometimes larger. The remainder of the pits use single bucket and wheel excavators with electric and diesel traction on rails.

Excavators with buckets of 12.5 m^3 capacity are in common use, while others of 20 m^3 content are in production. A scraper with 35 m^3 buckets and a jib 65 m's in length have passed the design stage. The manufacture and assembly of a drag line with bucket betweem 80 and 100 m^3 including a jib 100 m's in length has been completed. Over three thousand dump trucks of capacity varying up to 120 tons are used in open pits.

Hydraulic Mining

This was first carried out thirty years ago. In the Donetz and Kuznetsk basins where ten mines have been converted to hydraulic wholly or in part.

Soft coal is mainly recovered by hydraulic methods. The harder types of coal are mined by a combination of the two methods, although a move towards "all hydraulic" has increased over the past ten years through more efficient equipment, improved technology, the introduction of hydraulic transport and the dewatering of coal.

The main types of equipment used include:

1. Remote control hydraulic monitors GMDZ-3 and 4 rated at pressures of 12 and 16 MPa (equivalent to 1740 and 2318 psi) and using 100 m^3 of water per hour. It should be noted that MPa - Mega Pascal - is equivalent to 10^6 newtons per square metre where 0.69 newtons per square metre is equivalent to 1 psl.

2. The self-propelled programme involves the central hydraulic monitor 12 GP-2 rated at 12 MPa with a rate of discharge of water at 450 m^3 per hour.

3. The self-propelled hydraulic monitor with the pressure booster SGU-2M incorporated. This is supported by the lightweight mechanical hydraulic machine 1 MGP-5, to drive entries into steep seams. A mechanico-hydraulic machine MGPP-3A is used for shale, slate or sandstone. Both are fitted with hydraulic turbines approved as safe - even in gassy mines.

Coal slurry pumps which are effective and reliable,high pressure pumps, feeders and ancillary equipment have been developed for use with hydraulic systems. These include:

- a small one-stage coal slurry pump 12u-10- horizontal hydraulic transport systems rated for 900m^3 per hour at 90 mm's waterhead pressure.

- two stage, high pressure coal slurry pumps, 14 UV-6, designed for 900 to 1000 m^3 per hour through put at 320 to 330 mm's waterhead pressure. Also hydraulic lift and transport systems.

- section pumps 12 MSG-7 supply water for hydraulic monitors at a rate of 800 m^3 per hour with pressure up to 10 MPa (1450 psi) using water containing suspended solids at 70 gms per litre. It is also essential that pipe fitting is perfected together with couplings and other fittings while quick assembly and rate of dismantling are vital to control the amount of solids which are brought into the water in the system. Special valves 3 pp 300 have been designed for pressures of 6.4 MPa (928 psi) slurry pipelines and ore now in mass production - according to reports.

- instrumentation to control the solids concentration in open flow situation and pipelines have been developed.
- instrumentation consisting of a set of instruments (KPU) can now be used to record output from individual coal faces and sections of an entire mine including the ash content of the coal during flow and in flotation tailings.

Methods of Mining

Ideally this method should be used on softer coals, or those of medium hardness where monitors are capable of a faster rate of coal recovery than would be the case without their introduction.

Long pillar methods involving two headings 100 to 200 metres apart are driven into the seam for between 1.5 to 2 kilometres, when, as has already been indicated, recovery is by retreat mining.

However, considerable problems can arise during the use of powered supports or with the use of hydromonitors when floors break up as a result of poor contact between the powered support and the roof. Flumes are often used instead of conveyors which make recovery of between 7,000 and 8,000 tons mined in a single day from one face possible.

But these tonnages are dependent upon ideal conditions prevailing. At this point, it might be helpful to illustrate the two possibilities for incorporating hydraulic methods. This means the combined mechanical/hydraulic mining systems which use the series K-56 power loaders equipped with hydraulic monitors to help wash the coal away from the face. This method is illustrated in Fig. 6, while Fig. 7 shows the position of the power loader - magnified.

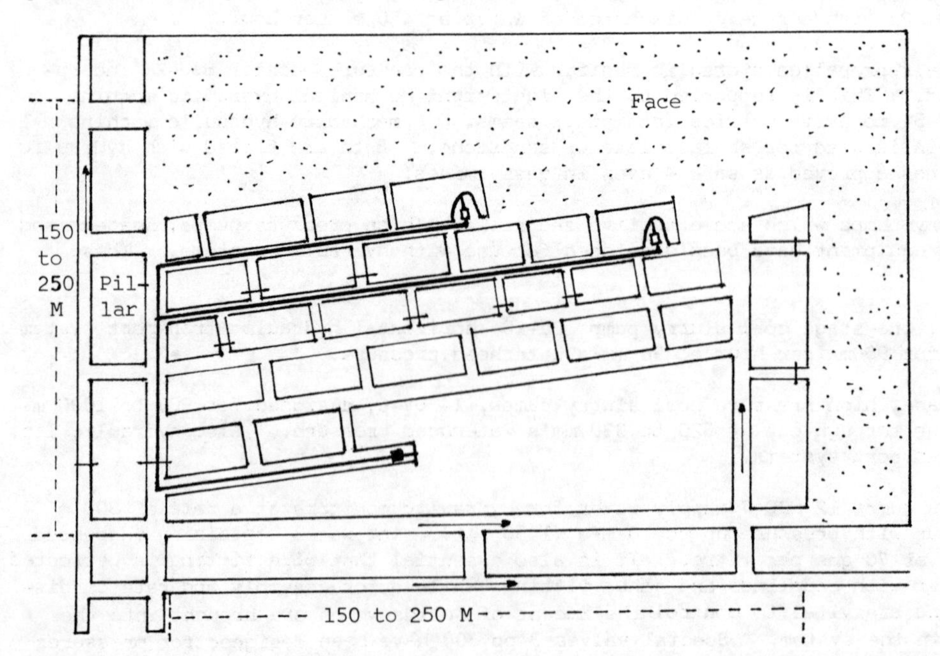

Fig. 6. Combined mechanical-hydraulic mining systems - USSR

Fig. 6 illustrates the manner in which the combined system is applied to work continuing simultaneously in more than one seam. One of the main benefits is virtually dust-free mining.

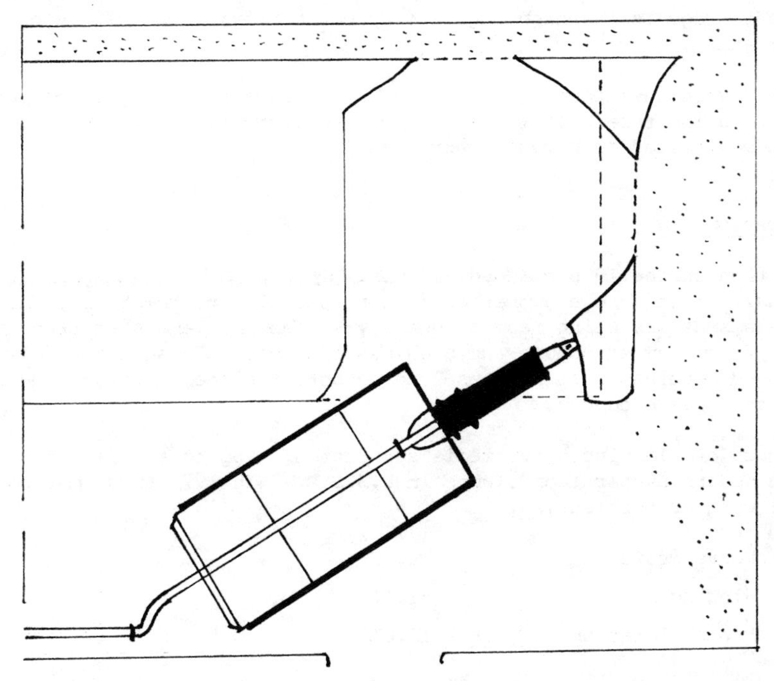

Fig. 7. The power loader at work - USSR

The Krasnogorskaya Mine. This provides a good example of the conditions under which hydraulic mining may function. Here the mine is contained by an anticline mined by the upper working levels. The coal seams are pitched steeply, under 50 to 80 degrees, with thickness varying from 1.5 to 9 metres. The formation contains a considerable number of faults. The coal is suitable for coking while the relative gas yield is reported as being between 9 and 30 cubic metres per ton.

Both development of the mine as well as the mining itself is carried out using high pressure hydraulic methods of cutting under the control of GMDZ-3M and 12GD monitors.

The basic method employed consists of sublevel hydraulic cutting, under flexible steel mesh lagging mounted in one plane limited by the walls of the seam at each fourth sublead. This flexible lagging is not used with seams of medium thickness.

Seams are opened up in blocks about 150 metres maximum length because of the contiguousness of the seams and the liability to spent combustion of the coal.

These blocks are opened up by intermediate crosscuts from the lateral drift driven parallel to the central line of the anticline, several blocks being in operation at the same time.

Within each block coal is moved by water contained in open flumes down to a coal slurry pump station to be found at the intersection of the lateral drift and the crosscut. Such a system, given the right geological conditions, reasonable roof and floor states, medium to soft coal and gradients favourable to flow from the face to shaft bottom, lack of dust will result and diseases connected with dust are abated.

A further advantage is the high degree of productivity; about 70 people producing 1.5 mts per annum.

Water in the hydromines permeates the coal seams partially dissolving the free methane and so reducing threshold levels. For the extraction of roadways, jets are either hand-controlled or remote-controlled.

Coal Preparation

Coal production during 1976 reached 712 mts. By 1980 this is expected to reach 800 million tons. There are a number of factors peculiar to the Soviet Union and yet increasing demands are being made on quality. The problems stem from mining of thin seams of complex structure and high-ash content. In addition bulk mining in coal faces and headings will increase the amount of stone, moisture and fines content of coals as produced.

Modern methods of cleaning,heavy media separation, jigging and flotation are all included in modern preparation plants in U.S.S.R. In 1975 these methods were used in the following proportions:

Heavy Media	24.4%
Jigging	51.1%
Froth Flotation	10.6%

These methods account in total for 86.1% of the total. Use is made of flow sheets, the one shown in Fig. 8 being that for a "run of the mine" coal. These include wet, dry and combined methods. In general, coking coals are washed down to 0 millemetres with steam coals down to 13 millemetres - sometimes down to 0 millemetres, depending upon technical and economic factors.

As can be seen from Fig. 8 the most widely spread flow sheets include heavy media separation or jigging of coarse coal and flotation of slimes. Occasionally dense media cyclones are used for washing fine coals in place of jigs. In U.S.S.R. it is compulsory to recover middlings or low energy coal.

Flow sheets for the cleaning of steam coals include heavy media separation, jigging, pneumatic and counter current separation as well as washing in spiral separators. Slimes are processed by flotation. It is claimed that the Komendantskoya Central Preparation Plant in the Donets basin is the first to introduce anywhere in the world the technology of the deepwashing of anthracite.

Equipment in common use or developed includes:

- The cylindrical screen GTsL with a special bell-shape screening surface and a capacity up to 100 tons per hour for primary sizing of coal.

- The GCL rinsing screen for classification of coal by sizes prior to washing. Capacity up to 800 tons per hour and intended for 13 millemetre sizing. Specific performance of the screen is 100 tons per hour per square metre with a water consumption of one cubic metre per ton of coal.

- an integrated system of equipment for treating 13-300 millemetres coal in the STT-type dense media separator

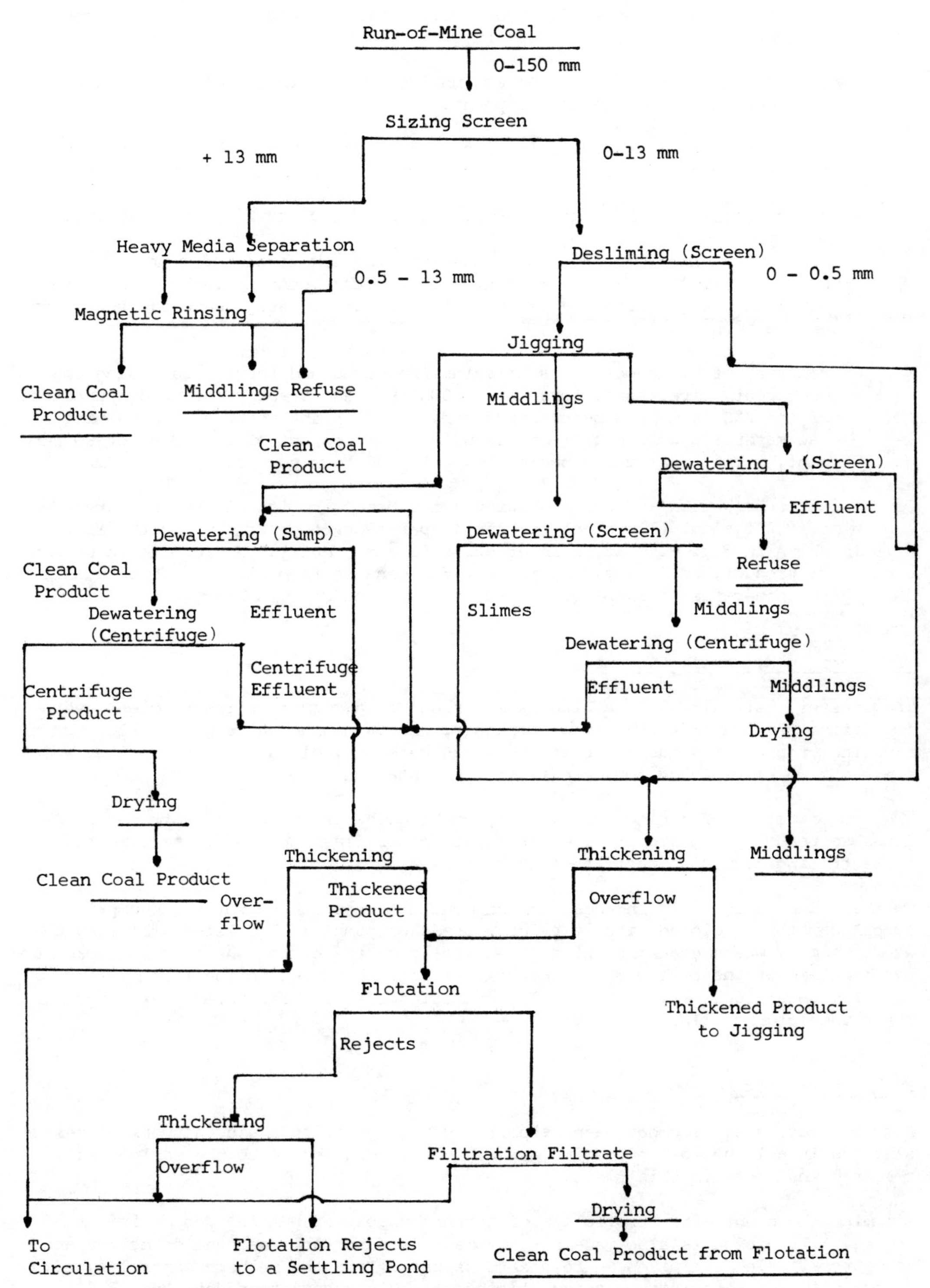

Fig. 8. Coal preparation flow sheet - USSR

- a three product dense-media hydrocyclone for separating coal into three products at the same media density.

- jigs for washing coarse, fine and unsized coal with capacities up to 750 tons per hour are under development.

- horizontal vibrating and pulsating centrifuge, VG8-2, has been developed for dewatering fine size clear coal produced during the jigging operation.

- vacuum filters with working surface areas of 140 and 250 square metres.

- a settling and filtering centrifuge ensuring dewatering of slimes and flotation concentrates with a reduction in moisture content in the final product equal to that produced by vacuum filters, particularly where the fines content is high.

- recovery of clean coal in submarginal reserves and impure coal using countercurrent spiral separators of the SSh type and steeply inclined separators of the KNS type have been introduced. This plant ensures good washing of raw materials of 150 millemetres with stone content of in excess of 80 per cent. The separator capacity is up to 400 tons per hour.

Processing of brown coal is carried out for the Moscow region at the Kimovsky open cast mine. A typical flow sheet for the preparation of brown coal in the Moscow area is shown in Fig. 9. It is interesting to note that sulphuric acid is obtained as a bi-product, while the clay is sent for construction sites. These plants satisfy the non-refuse technological processes in the coal industry.

Transport in Hydraulic Systems

The present systems result in high water useage as a result of insufficient hydraulic cutting particularly in drivage workings, and from high water consumption for the flushing of pipelines due to great distances between individual working sites and the short working period for each slurry pump station.

The concentration of mining effort into one block has resulted in shortened pipelines so reducing the amount of water required for flushing and at the same time concentrating transport facilities in one place.

On the other hand, it is intended to use monitors with a high water consumption supplied within a closed circuit by high pressure pumps. The water will accumulate in sumps, where coal precipitates and falls to the bottom where it is picked up by an intake of the coal pump. The mixture is then pumped to the main lifting sump for onward delivery to the coal preparation plant. The water is returned for reuse after cooling.

Health and Safety

A mobile rescue service has been established incorporating a wide network of rescue stations in all the coal producing areas. It is claimed to be the world's only research institute in this field.

All mines have anti-fire protection equipment capable of putting out a fire at source. Provision is also made for automatic and portable methane detectors and indicators. In most mines where hazardous gas exists the methane control systems switch off the electricity when quantities of the gas have been detected.

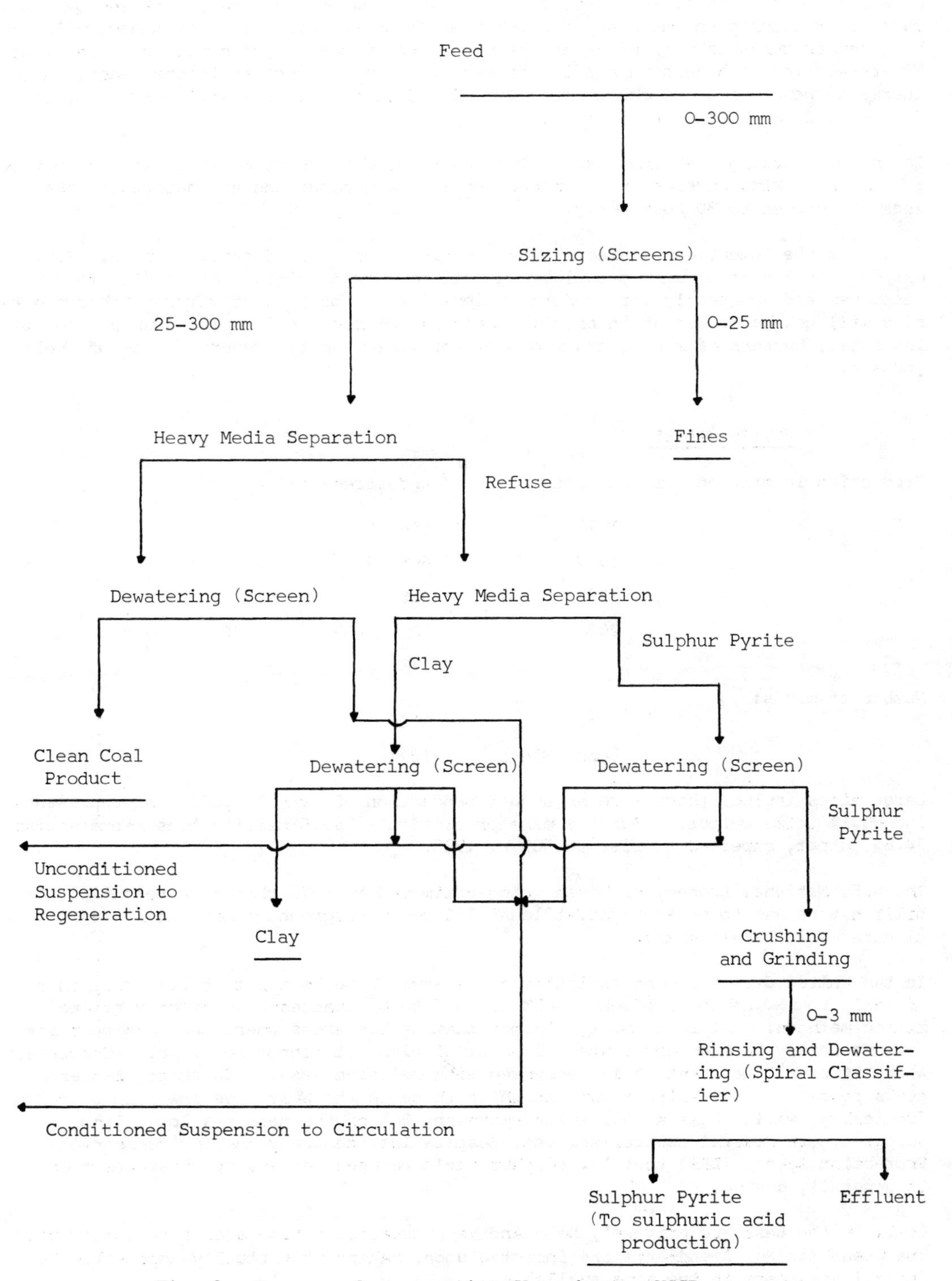

Fig. 9. Brown coal preparation flow sheet - USSR

There is a programme to bring about a reduction of harmful dust which causes mining diseases and also to prevent coal dust explosions as well as coal, rock and gas outbursts. Rockburst and methods of shock prevention are also being examined,while high temperatures in deep mines are reduced through the use of mobile cooling plants. Miners working in hazardous conditions received ten per cent additional wages while increased pensions go to those with over 15 and also over 20 years' service of 140 Roubles and 160 Roubles respectively.

The normal working week amounts to 41 hours reduced to 36 hours with two days off for those working underground. Where conditions underground are hazardous, the week is reduced to 30 hours only.

It is not the intention to examine every mining country in detail. The U.S.S.R. has been looked at in some detail because it is a major mining nation with vast resources and apparently very modern equipment and good productivity. Other countries will now be looked at in brief, particular emphasis being placed upon production rates, numbers of mines, types of coal produced and the general state of their industry.

UNITED STATES

Production in millions of tons both achieved and forecast:

1975	648
1977	665
1985	956
2000	12 - 1400

Number of mines:

Total	6168
Large Mines	1074

Large mines include those with an annual production of over 100,000 tons representing 80.1% total output. Surface mine production - 355.6 million tons representing 54.8% output, expected to rise to 60% by 1985.

The U.S. National Academy of Engineering estimated in 1975 that if a target of 1.2 billion tons was to be achieved,at least 140 new underground mines and 130 additional pits would be necessary.

In the United States recent estimates of reserves have been put at 3.2 trillion tons of coal, so far unmined, of which 85% is said to be inaccessible by conventional mining methods. It is commonly claimed that in the areas where coal can be burnt satisfactorily from an environmental point of view, it cannot be mined. Conversely where it cannot be burnt,it is recovered with relative ease. In short, Eastern coals possess a high sulphur content while those in the West have low sulphur values. Ironically, while these areas, which represent 80% of the nation's low sulphur supplies, are Federal controlled, yet, despite insistence by the Environmental Protection Agency (EPA) that low sulphur fuels be used, access to these reserves is generally denied.

Coals in the East are privately held and yet unrealistic time scales to comply with the Clean Air Act Amendments are insisted upon, making it virtually impossible to install scrubbers in the time available.

Mining Methods

Longwall mining is not always feasible or practical. At present, it represents 3% only of total mining activity. Shortwall mining machines are at present filling the gap between longwall mining and the continuous method of mining.

Streamline mining machines are replacing large entry filling machines, while more room now exists within the entry so that roof supports and ventilation of the face may be achieved. Much less movement of the continuous miner from the working area now takes place.

Transport Underground

The greatest need here is greater reliability. The problem of bringing up the continuous face haulage remains. Continuous haulage is absolutely vital. Shuttle cars are used in more than 75% of coal mines.

Surface Mining

This is carried out in most strip mining situations using mainly walking dragline machines.

CHINA

Details as to numbers of pits and capacities do not seem so easy to come by, China's coal industry being affected by political considerations. The latter, coupled with the natural disasters of 1976, caused very considerable disruption. It was not therefore until 1977 that the Ministry of Coal,re-established in 1975, was able to make its presence felt and get its 10 year programme underway.

Tonnages produced in millions of tons, fact and forecast, are:

1976	480
1977	500 to 535
1980	600

In China, coal is the main source of energy today, although it is expected that this pattern will change in this way:

1977	67%
1980	60%
1985	50%

China's reserves of coal include 35% of coking types said to be of poor quality, of which 11.2% are in the north with 18.9% in the northwest. The northern province of Shansi, together with Hopeh, Honan and Shensi, possess 70% of these reserves, while other large deposits are to be found in Heilung Kiong, Kirin, Liaoning and parts of Szechwan. This uneven distribution of reserves creates considerable transport problems since factories and other large users are mainly in the south. However, production in the south is expanding twice as fast as in the north; for instance, in 1975, 31 new pits were opened with capacities of over 100,000 tons.

Open Cast. This amounts to 10% of total coal production.

Meiyukou Mine

The equipment in this mine provides a useful guide as to the types of equipment being used. This is now produced by over 100 factories supported by a large number of repair centres.

This mine produced 1.6 mts of the output from the Shansi Provinces. It has four shafts, two vertical and two inclined. The 4.5 metre diameter inclined service shaft is powered by a 1000 hp a.c. winder.

Seams are found between 1 and 10 metres thick. There are 11 faces mined by the longwall approach and using the method of retreat mining. 3,out of the 11 faces, produce the major quantity of the coal using 2½ shifts with an output of 1000 tons per shift. Shearer Loaders capable of a 60 centimetre cut at a single pass are employed, the coal being removed on to a conveyor belt 3 metres wide.

Face supports are of the Chinese Tatung EB1-400 chocks type fitted with hydraulic valves. The canopy has an area of 5.2 sq. metres exerting 31.5 tons pressure. The conveyor rams employ 15.4 tons pressure. The chock advance ram which exerts 9.7 tons pressure has a range of 0.92 to 1.7 metres using a 5% oil to water SRMI pump at 1000 kilograms of fluid per minute.

The face conveyor is a 0.6 metre Polish model double drum shearer both cutting 0.3 metres, accompanied by a face advance of 0.6 metres, while the drum diameter is 1.6 metres. The shearer runs at 1000 volts. The pit employs 2,500 people below ground. There are plans for increased production of 12.5% to 1.8 mts by 1978, 2.2 mts by 1980 and 2.7 mts by 1985. The next major investment is to be in improved longwall equipment.

Open Cast Mining

In Liaoning there is the Fushun West Open-Pit Mine with an annual output of 15 mts. It covers an area 6.6 kilometres in length and is 2 kilometres wide with a depth of 260 metres being used to develop a single seam varying in thickness between 40 and 120 metres, but averaging 80 metres. Development proceeds through blast holes being drilled 4.5 metres apart. Each charge consists of 8 kilograms of ammonium nitrate. The holes are drilled with 160 millimetre fish-tailed bits, rigged for tungsten carbide inserts, which must be resharpened after every third hole.

The overburden is removed by shovels of 4 cubic metres capacity loaded on to 14-car trains consisting of 4 ton waggons serviced by ten 1.5 to 3.0 cubic metre electric shovels. Locomotives operating at 500 volts, direct current, run on 900 millimetre gauge track, haul the trains. The ratio of overburden to coal is said to be 6.4:1. With a 16,000 workforce, plans are afoot to deepen the mine to 500 metres.

Coal Preparation

This important aspect of the marketing of coal leaves a great deal to be desired, particularly with exports to Japan in mind. Much better facilities for moving coal to the ports are also vital. It also stresses the necessity for installing mining techniques which are cleaner and more efficient.

Future Needs

The main mining requirements, for the future, must, short-term, be imported. These include:

- coal drilling and preparation equipment
- large capacity trucks and front-end loaders
- shovels for open pit mining
- face extraction systems, self-advancing hydraulic roofs and supports
- shearers, conveyors, lighting and communication equipment.

Coal production in China is a major element in the Grand Plan designed to overtake the United States by the end of the century. But much will need to be done to achieve this goal.

INDIA

Production of coal, in millions of tons, is expected to follow this pattern:

1977	100
1985	200

But in order to achieve the 1980 target, some 60 new mines will be needed, particularly extensive mechanization for both surface and deep mines. The equipment is expected to be produced locally with the mining machinery being imported.

Total reserves amount to 23,139 million tons of coal of which 10,643 mts are believed to be recoverable economically in respect of hard coals and 897 mts in respect of lower grades.

POLAND

The present position with regard to coal production with forecasts to the year 2000, in millions of tons,is:

1977	170
1985	215
2000	250

Coal reserves are rich, with easy access. In the upper Silesian Basins alone deposits of coal are said to go down to 1000 metres totalling 75 billion tons with 16 billion tons being coking coal. In the Lublin Basin reserves total 40 billion tons.

The Jan mine is the first mine to be completely automated. It is claimed to be the most advanced mine, being ahead of world technology. This incorporates an automated isotope face system for mining coal in the absence of men. Even so, the mine employs 250 in total.

AUSTRALIA

Present production is expected to rise so far as black coal is concerned by over 50% over the next 20 years, figures in millions of tons amounting to:

1976	85 black	30 brown
2000	130 black	

Figures in respect of brown coal needs for 2000 are less certain but are unlikely to increase by as much in proportion.

One of the major problems in Australian mining is the raising of capital, 51% Australian holding being the figure envisaged in the New South Wales guidelines an "investment in the new mineral exploration and development in the State", although a gradual build-up to an Australian majority is permitted. For instance, British Petroleum have a link with Oakbridge to build a new mine at Clarence.

Australian reserves of coal total 74341 million tons. Of those considered to be recoverable, 13,770 million tons are accounted for by hard coals with 10,555 million tons of lower ranked types.

Queensland, New South Wales and Victoria are all states with a promising future in coal, Queensland in particular possessing large capacity stripmines which are being developed.

In New South Wales mechanized room and pillar working will be used with longwall employed on a limited scale only, where thick seams are involved. Shield type self-advancing roof supports could well have an application here - according to roof conditions. Longwall and shortwall methods, both employing retreat mining could well be used for a considerable part of the anticipated increase in annual output. The needs of Japanese steel mills for increased quantities of coking coals amounting to 50 million tons will be a major reason for the predicted coal development.

SOUTH AFRICA

Coal production and forecasts in millions of tons have been set at:

1977	76
1980	150
2000	236

Backing these large increases in production are reserves estimated in millions of tons at:

Total Reserves	44,339
Economically Recoverable	10,584

These quantities compare with figures for the whole African continent of 59,000 mts and 15,000 mts respectively. South Africa holds 75% on each count.

Mining Methods

Horizontal seams of shallow depth prevail and therefore room and pillar systems are used, with the pillars being left unworked due to the legal position regarding subsidence. There is no right to let down the surface of the ground. As a result, recovery only just exceeds 50%. Further, the price ceiling limits the degree of extraction.

Coals are very tough to cut which does not lead to continuous miners with the ripper type cutting heads. It is therefore vital that more intense methods are employed.

At present, Arcwall cutting machines and coal blasting methods are used, after which "gathering-arms" collect the coal and place it into shuttle cars.

The tendency towards increased labour costs will lead to open pit mining where restoration of the surface is much easier.

Bosjessprunt. An underground mine providing the total requirements for South Africa's second oil from coal plant, SASOL 11; it has an annual capacity of 12 mts.

Kleinkopje. This is the latest opencast mine with an annual output of 5.5 mts by the 1980's. A Bucyrus-Eric dragline for overburden removal with bucket capacities of 34.4 cubic metres is being used.

Health and Safety

The Chamber of Mines of South Africa are responsible for mine safety. This includes a mine rescue system involving two large Ingersoll Rand drills and a 5,000 cubic foot per minute 125 psi gauge portable compressor. One drill is a 6½ inch pilot drill, while the other involves a large 30" diameter type. The former is used for air, food and small items, while the large drill hole takes the men. Fortunately the accident rate is termed as favourable.

CANADA

Production of coal fell between 1975 and 1977 as the figures, measured in millions of tons, show, but a large increase is anticipated by 2000:

1975	27.8
1977	20
2000	60 to 75

Reserves are claimed to total 120,237 million tons, although other sources would place them rather lower, with hard coal reserves, which are economically recoverable, put at 4,195 mts.

These reserves stretch across Canada, with 30 mines operating in five provinces plus one small mine in the Yukon.

Government policy and incentives play a considerable part as has been seen in other countries, with the Federal Government evolving a "coal policy" to overcome the problem of provincial ownership of minerals.

A number of mines may be used to illustrate the state of the art in Canada.

Nova Scotia. This State boasts three longwall faces with a daily output of 8,200 tons.

Alberta. Here failure in longwall mining has been the experience, but fresh attempts are expected in Western Canada, the main sphere of increase in mining activity.

East British Columbia. This area is at present practising hydraulic mining above the drainage level which is soon to be extended below it. Koiser Resources also have a large new hydraulic mine with a production target of 2 million tons by 1980.

South East British Columbia. Here, an underground hydraulic mine is believed to reach 1 million tons during 1978. Seams in this mine are thick - between 25 and 30 feet - at a gradient of 4°. All mining occurs above the natural drainage level.

The Prince Mine. This working has three production panels operating continuous miners and shuttle cars, scheduled to reach 800 tons daily per panel by 2000.

Stripmining

Massive surface mining efforts are carried out, 5.6 million tons of clean coal being produced.

EUROPEAN COMMUNITY

There are five coal producing countries within the European Economic Community which in 1977 - in millions of tons - produced:

United Kingdom	120.7
Federal German Republic	91.2
France	21.3
Belgium	7.0
Italy	2.0
	242.2

Although not yet members of the community, but likely to join, Spain at 11 million tons produces more coal than Belgium or Italy. Spain's energy needs required 15.4 million tons of coal in 1975, although production, as has already been indicated, was lower than this figure, with a target for 1985 of 26.6 mts. And yet the coking coal requirements are expected to reach 35.3 million tons by 1985.

Coal requirements for the Community as a whole - that is the Nine less Spain - are expected to be at around 220 mts as indicated in Table 3. The two main suppliers are clearly the United Kingdom and the Federal Republic of Germany. The latter visualizes little growth for the time being mainly due to a fall off in both electricity demand and steel, applying to black and brown coal. However, a long term need for between 105 and 155 mts is seen as being necessary plus a further 85 mts for replacement of natural gas by synthetic natural gas from coal. The immediate problem is to maintain present levels of hard coal at 85 mts down from 150 mts in 1957. European methods will be examined through the eyes of the United Kingdom coal industry, although Germany has a massive research and development programme into coal which will be looked at later. In Germany the cost of labour is very high indeed: research is seen as the way out.

UNITED KINGDOM

Historical and current rates of production of coal together with future targets, in millions of tons, are seen as:

1973	133
1977	120.7
1985	150

Open Cast output in 1976/7 totalled 11.25 mts. Reserves of coal have recently been reviewed by the National Coal Board's Central Planning Unit - 1978. Then it was announced that economically recoverable reserves totalled 637,000 mts estimated to be sufficient for 235 years. Ultimately recoverable reserves will probably total 1000 billion tons which at the current rate of recovery would last for 370 years.

Coal Reserves under the Sea

Now that coal exploration has been extended to areas covered by sea, an interesting situation arises. The United Nations Law of the Sea Conference has been attempting to define international regime for ocean mining. The concept of resources of the ocean floor has always been one of common heritage, but the realities are that the technology for exploitation is narrowly held, the influence of the United States being dominant and resulting in a conflict of international sovereignty versus commercial considerations. This raises general principles involving rich and poor nations, those with coastal areas such as U.K. and those which are land locked - the Haves and the Have Nots.

Mines

There are 231 mines in the United Kingdom run by the National Coal Board (NCB) producing deep mined coal, but many are small, the bulk of the coal coming from the few. New exploration and development has been concentrated in three areas.

- North of Selby, 10 million tons per annum.
- Vale of Belvoir and the area between Newark and Lincoln.
- Extension of existing Midlands mines around Park in Staffordshire

It is from these areas that the additional 30 million tons by 1985 will come, together with increased output from existing collieries.

Mining

The perfecting of mechanized longwall methods of mining is one of the NCB's main objectives. In 1974 there were 750 longwall faces rising to 760 during 1975 but falling back again to 730 by the end of that year. Daily face output was recorded at 605 tons. The technique is well established and has proved itself capable of producing very high outputs under favourable conditions. In the United States, as has already been discussed, it has been considered superior to their own long established bord-and-pillar methods, where geological conditions approach those encountered in the U.K. This has resulted in British manufacturers of mining equipment experiencing considerable success there, with equipment based on that usually used in Britain.

Changes stemming from increases in resources employed and methods of mining have resulted in great development of drivages.

One of these changes has been the application of retreat mining - that is beginning at a point distant from the roadway and working back to it. This has created the necessity for increased yards of drivage per 1000 tons of output.

Despite a rise in the number of heading machines from about 250 in 1970 to almost 600 in 1975, the average daily advance has not increased, and yet, an additional number of man-shifts per drivage shift from under 8 in 1970 to 12 during 1975 has taken place.

Retreat working which permits a marked simplification in the mining system increased during 1977 to produce over 20% of the total major longwall face output. This method of retreat mining shows an increased output over advance mining methods. But it is limited in scope by the rate at which drivages to open up the retreat panels can be made.

Much effort has gone into improving operations at the face ends of advancing faces so as to bring the potential nearer to that for retreat mining. The number of faces using ripping machines continued to increase as did the number with mechanized packing and those with heading type systems. In particular, the recently developed NCB/Doseo in-seam-miner has done much to eliminate stables and the working of face ends. There are 120 already in operation in U.K. pits.

Haulage. A switch to chainless power loaders made the number in use virtually double during 1976/7. Seven types are at present in use and under test, but the principle has been fully established as being safer, quieter and as a method of haulage more reliable, than the static chain previously used. The system is expected to expand rapidly.

The increased weight and greater capacity of power loaders has created a need for heavy duty face conveyors and led to a change in design.

However, in the search for greater tonnages of coal, and in turn increasing lengths of coal face, it should not be forgotten that there is a statutory limit in the United Kingdom of two metres on the distance between the coal face and the first line of supports. This restricts increases in the depth of the web. Major increases, however, would require a relaxation in the limits before an increase in the prop-free-front distance can be brought about.

Roof Coal. This additional reserve has always been a feature of British mining, that is the coal left in the roof to improve strata control and as a result of difficulties in extracting very thick seams.

Chock/shield types of powered supports will assist this process of more complete coal removal. Otherwise there are 12,800 powered supports in use with 79,000 more being installed or reconditioned.

Shearers. These are specially designed for thin seams and increasing productivity in medium seams, where they are known as floor-based shearers. The aim now is to reduce their cross-section and fit chainless haulage. The NCB has 1000 shearers of different types with revolving drums varying from 30 to 72 inches diameter powered by 100 to 400 H.P. motors and weighing up to 23 tons.

Conveyors. Used extensively, the world's largest belt being sited at the Longannet drift mine in Scotland where it extends 5½ miles underground linked to a two-way radio control system.

Coal Preparation. This important above ground activity is to be seen in its most modern form at the near completed plant at the Prince of Wales Mine at Pontefract. Coal is fed into a hopper from which it is delivered on to a crusher after which it is passed over a screener. The large coal is either recycled or taken off the screen and used in the larger form.

The fine coal then passes into a water separator after being mixed with magnetite which increases the density of the slurry. The latter then falls to the bottom of the separator and is removed, while the coal rises and is taken off. The magnetite is recovered because being an ore it is expensive to mine and therefore too valueable to throw away.

The coal is now passed over a vacuum filter while the shale is filtered out separately and disposed of. The whole system has been designed to use minimum energy demands.

Opencast

The NCB have experienced increasing difficulty in gaining public acceptance of opencast mining, particularly when coal stocks are rising. This source of coal represents an essential means of obtaining certain coking and naturally smokeless coals. Coupled with low production costs and a reduction in imports, this type of mining provides a natural incentive to increase production to at least 15 million tons per annum.

Health and Safety

Significant quantities of methane are drained from under-ground mines for safety reasons. More and more of this gas is being used either at the colliery or is sold to third parties. During 1976/7 the quantity of methane used in this way increased by 40%. It included gas firing of boilers at five collieries and one coking plant as well as a Distillery. One limiting factor is the requirement for a 30% flammable gas content as is normal in most overseas countries as opposed to the U.K. legal limit of 40%. It is hoped that the lower figure may be permitted.

In 1976/7 there were 38 fatal accidents, 21 down on the previous year, while serious injuries were also down at 538 from 515, continuing a down trend which has continued for the past few years. An improved scheme for training men in the use of self-rescue was introduced. Particular precautions to avoid accidents are taken, including the banning of electric watches underground, while all hydraulic equipment used underground must be adapted to use an emulsion containing a low proportion of oil in water as opposed to pure mineral oil used in surface equipment, which is inflammable. Work has continued to reduce dust made by increasing pick penetration to provide a coarser product. Heavy duty cutting tools and shearer drums rotating at lower speeds increase this pick penetration, made possible by an increase of 14% in the number fitted with two speed gearboxes and heavy duty picks.

1976 saw the lowest percentage of certifiable pneumoconiosis cases ever, at 2%, from the 51,000 miners X-rayed.

Training

A new National Scheme of Training for Work for Coal Production was approved by the Health and Safety Executive, the controlling body in U.K. This includes recognition of coal drivages as a proper place for some periods for training recruits to the mining industry and others.

Environment

Environmental matters are central to the industry's main activities, made more difficult by the concentration of population around the coal fields coupled with a very high ratio of overburden to coal as compared with opencast operations abroad.

In particular, the land used in opencast is restored to the highest possible standards after mining has ceased. Air pollution is controlled through efficient use of coal and by requiring large boilers to be used in conjunction with chimneys sufficiently tall to provide full dispersal of all emissions so ensuring that concentrations of sulphur dioxide and other pollutants emitted do not constitute a health risk.

New Mines. Figuring high up in environmental consideration, special provision is being made for methods of washing, waste disposal and coal despatch; to minimize disturbance. At Selby for instance the winding mechanism was restricted as to the height which should be viable above ground. Restrictions were also placed as to mining activity permissable immediately below the vicinity of the Cathedral and within close proximity of the river, to minimize the risk of subsidence, particularly in view of the fact that much of the surrounding land is at or below sea level.

REALITIES OF LIFE

Clearly the descriptions which have gone before cannot all represent personal experience. Much of the information has been taken from reports and accounts which others have given. High standards are relative: what is considered to be excellence in one country may be better or worse than excellence in another. The same reasoning must apply to comparisons in output. Those countries with the greatest amount of mechanization probably produce the greatest amount of coal as a result, but it could also be that safety takes on a lesser role in one country than in another, where relative output is as high. Equally uncertain labour relations, as during strikes in Britain in 1974 and the United States in 1978, were the prime cause of poor output.

Those who have been to the U.S.S.R. will have viewed the bald tyres on the aircraft used for internal flights with considerable cause for concern. It would not be unreasonable to suppose that the difference in standards in this field as compared with Europe might be repeated in the mines, despite a programme apparently aimed at high standards of safety underground.

Similarly, engineers from Europe who have visited Polish mines tell of temperatures, even heat, so intense as to make working underground unsafe. Heat in the mines is also experienced in Germany.

Equally the degree of mechanization may be tied to labour rates. Where wages are high as in Germany, investment in capital equipment is a prime objective. On the other hand, in India, where labour is plentiful and very cheap, jobs are more important than investment, much of which will be in equipment from overseas, resulting in an outflow of foreign reserves.

Chapter 4

NATIONAL RESEARCH AND DEVELOPMENT PROGRAMMES

Mining is dangerous; it always has been dangerous, but then so is deepsea fishing, piloting an aircraft, working on infectious diseases or being a steeplejack. As an insurance risk mining today does not head the list for loaded premiums.

Even so, much of the inherent risk can be reduced by continuing research into health problems and the development of safer equipment and methods of recovering coal. Dust has always been the miner's main enemy, from the point of view of explosion, but particularly because it has caused such havoc with health over the years. As can be seen from the British experience the incidence of pneumoconiosis is falling rapidly as judged by annual X-rays, but even one new case is too many. Success really hinges upon keeping down dust and paying particular attention to the design of the equipment which creates it.

Gas, another killer, must be detected quickly, but better still it should be lead away for use as a fuel.

Nature has presented the coal miner with natural obstacles to the removal of coal. Geological variations, changes in directions and level of seams, their constituent chemical content and the thickness of the seam all present challenges to research and development. But coal must be useable when presented to the consumer. It must be the right size, have the right ash and moisture content and give off volatiles appropriate to the areas in which it is being burnt.

This provides problems for the designer of the preparation plant, on the one hand, and also the manufacturers of equipment on the other. But the object of this chapter is to review R&D into mining equipment and its impact upon those who work in the industry and those who use the coal or are likely to be affected by its use. R&D into conversion of coal will be dealt with in Chapter 9 where progress in liquefaction and gasification technology will be examined together with more efficient methods of combustion. Returning to conventional mining, it soon becomes clear that considerable work is taking place in all countries where mining is practised, although it will not be possible to examine every country here.

U.S.S.R.

The objectives in the U.S.S.R. are without doubt the same as for all other mining countries - to increase efficiency, make mining less arduous and create safer working conditions all to be accompanied by high wages and a 30 hour week.

Research Organizations

The development of coal fields will be influenced by their geographical location from the Arctic Regions down to Central Asia and the Carpathians to Isle of Sakhalin. Within these bounds all the problems which R&D must overcome include permafrost, low and high temperatures, seismic activity, the prevention of rock bumps and sudden outbursts of coal and gas. Examination is covered by 36 specialized R&D institutes sponsored by the Ministry of Coal Industry, 30 Institutes of the Academy of Sciences plus 37 institutes operating under Ministry of Higher and Secondary Education, 66 research organizations and 86 specialized machine building plants.

Progress

Expenditure has risen from 17 million roubles in 1970 to 30 million in 1975 which is indicative of the rise, in real terms, of resources devoted to development, bearing in mind that in the U.S.S.R. non-convertability of the currency has contained inflation.

The main effect has been concentrated on sophisticated equipment. Mechanized complexes have more than doubled in number. Head and tunnelling half as much again, while the use of scientific instruments for automatic control has risen by 1.7 times.

Mines bearing 3 to 12 million tons of coal with strikes extending from 6 to 15 kilometres have been studied with the conclusion that the best method of mining lies in the block system with a single level mine layout being the most practical for flat seams - up to 12°.

Rockbreaking

The development of efficient techniques for rockbreaking has continued, where the aim is to increase the speed of drivage of roadways and the degree of mineral recovery. Mechanical breaking techniques are being improved by developing combustion methods using the low tear resistance of rocks and arriving at the most suitable conditions. An active cutting head, pulse rockbreaking and controlled drive are being employed. High torque hydromonitors have also been developed, which automatically control breaking parameters and also eliminate mechanical gears. They result in a very substantial reduction in the dynamic contribution of forces in the transmitting elements and the frequency of oscillations falls by up to 20% too, in comparision with a synchronous motor. The K128P power leader equipped with 450 killowatts drive is at an advanced trial stage.

Development of high performance cutting heads for heading machines is based upon this experience, together with impact breaking. A boom type heading machine, type 4PP2, has been developed for drivages in coal and rock of considerable hardness, while rotary-type heading machines with diesel cutting heads for drivages in harder rock are also well advanced in development.

Heading complexes, designed to increase the efficiency of roadway drivage, are being developed. They integrate a heading machine, arch support erection, drilling and a roof bolting unit, a stage leader, a telescopic belt conveyor and monorail, while some complexes include shuttle cars.

Seams

Seams between 0.7 and 1.2 metres thick employ mechanized complexes of what is known as the Donbass type and the MK97 type incorporating shearers and plows. Flat seams less than 0.8 metres thick result in complications, if complexes are used, and therefore underground auger mining is used with success, particularly in the Lvou-volyn basin. Complete mechanization, although making considerable progress in steep seams, accounts for 7 per cent only. The greater effect of gravity and faults, more sophisticated methane drainage systems and occasional floor instability all contribute to slow progress as compared with mechanization on the flat.

The KSBD complex has been developed in conjunction with experts from Bulgaria for use in mining steep seams of thickness varying from 2.6 to 6 metres and with dips from 40° to 80°. It includes remote controlled drilling and shot-firing methods with automatic loading protected by roadway roof supports. This removes men from the coal face.

One of the main means of technical progress in underground mining is the move to technology producing a small number of operations at the long wall face, resulting in a rise in productivity of between 200 and 400 per cent.

Open Cast

The development of transport technology has been influenced greatly by the introduction of drag lines with bucket capacities of 15 and 80 cubic metres and shovels with bucket capacities of 35 cubic metres. In addition, Bucket Wheel Excavators with capacities up to 12,500 cubic metres per hour and drag lines with buckets capable of holding 100 cubic metres, also dump trucks and coal haulers of up to between 180 and 240 tons are being developed.

Roof Supports

The aim underground is for complexes with an ability to support the roof in conditions between 35 and 40 tons per square metre. This may need to be increased to between 70 and 140 tons per square metre depending upon the seam thickness. In some seams with difficult roofs and 30% of the faces totally mechanised artificial loosening of the roof is brought about by explosives.

Mining Hazards

Increased depth of mining brings problems such as sudden outbursts of gas, coal and stone, rock bumps and rock swelling, all presenting very considerable hazards This resulted in an integrated programme to prevent such outbursts, being set up by the specialist organizations. As a result, these dangers have been greatly reduced mainly through the working out of seams above and below tnat in question. One major advance has been the introduction of seismo-acoustic forecasting of the liability to sudden outbursts. Continuous recording of the intensity of the fragile breaking of coal seams and rocks takes place in 42 mines. Long term observations have shown that seismo-acoustic apparatus of ZUA-3 type gives warning from 6 to 20 hours before an outburst commences. At 160 mines, degasification of coal seams by thin seam satellites produce 2.5 million cubic metres of methane daily, which come up to the surface. It has also been necessary to develop a central system and automatic gas protection which eliminates the possibility of disturbing the degree of the ventilation at the coal face. The AMT-3 type of automatic gas protection apparatus has been developed and also introduced at coal mines. It deals with gassy coal seams as a result of which improved safety has resulted.

UNITED STATES

For many years, output increased regularly, that is up to around 1970, when output per man plummeted,many deep mines producing up to 35% less than formerly. This decline was generally put down to the method of enforcement of the Coal Mine Health and Safety Act.

If the industry is to recover, certain problems need to be dealt with. These are to:

- eliminate counter productive legislation based on emotion rather than logic.

- provide loan guarantees to the industry, in general, for development of better productive methods.

- arrange a coal export incentive programme to recover the position with regard to metallurgical coal as a principle export.

- to ensure price stabilization and to ensure reduced capital risk, also depreciation and amortisation - a vital factor.

The basic problem in 1975 was that little had been done by way of research since the days of the pick and shovel. That is the view of an American in the coal business. It was clear that a new concept of underground mining was vital. But it had to include additional safety for men, but at the same time reduce mainten-sance problems. In particular, a new method of transportation of coal was vital to remove the coal from the area of mining. This has led to the development of the movement of coal by hydraulic means. The Consolidated Coal Company are at present engaged on this aspect of development underground. The main problem has always been one of roof control - an area in which over the years many fatalities have taken place.

Research carried out in 1974 showed that tests carried out underground by injecting resin - low viscosity, non shrinking epoxy resin systems can be used for bonding the rock, thereby consolidating, stablizing and strengthening the roof structures. Injection under extreme pressure is necessary.

Longwall Mining

The problem of roof control applies particularly to this very safe method of mining. Other problems to be solved include materials handling, ventilation and general cyclical problems. Automation is the main objective to make longwall mining, which is much more continuous than the room and pillar method, less dependent upon men on the spot. It will also improve the quality of the product and worker safety.

No major changes in basic production are visualized for the immediate future. During the next ten years continuous mining, particularly longwall and shortwall systems will progress, while traditional methods will decline. Manufacturers of machines and equipment used in mining face the effects of legislation and unrealistic regulations, on air pollution, environment and other activities related to coal. As a result, engineering personnel are occupied designing modifications and accessories to present generation machines to comply with the Mine Enforcement Safety Administration (MESA) and the Environmental Protection Agency (EPA). Little time and money are therefore left for research into innovative machinery. A further brake upon progress is the degree of labour unrest, culminating in the strike which last for over three months.

Health and Safety

Research by the Bureau of Mines has shown clearly that under certain conditions methane can be drained from coal seams before mining begins, or removed from mined out areas, after mining, to prevent dangerous accumulations. Commercial techniques for the removal of methane have been developed now. This source of gas could make its own contribution to dwindling gas supplies.

The National Mine Health and Safety Academy - West Virginia - are in the process of designing, developing and conducting training programmes for mine inspectors to assist the Mine Enforcement and Safety Administration's efforts to reduce accidents and health hazards in mineral industries. Courses in mining safety, industrial psychology, hygiene and investigatory procedures have been set up.

Preparation of Coal

Two methods of arriving at improved quality have been developed.

The Catalytic Inc. and Syracuse Research Corporation have successfully produced checmical comminution fractures of coal along natural fault planes, exposing entrained inorganic sulphur and ash. Subsequent cleaning operations separate the coal from useless mineral matter. This process replaces mechanical crushing and grinding systems for size reduction. One hundred pounds of chemical comminutive are necessary to treat one ton of coal, 99% of which is recoverable.

The Massachussets Institute of Technology have been successful in removing substantially all the inorganic sulphur and forty per cent of the ash-forming materials from liquefied coal.

In 1974 a five year funding pattern programme was set up in which progressively increased sums were made available for coal extraction up to 1979. The allocation has been updated by Congress. The sums involved exceed those allocated by Congress to what was originally the Energy Research and Development Agency (ERDA) before the Department of Energy was created. This part of the programme covers the non-mining aspect as shown in Table 14.

This programme is divided into two parts. Near term, that is up to 1985, Low Btu Gas will be the main objective. The remainder, particularly High and Medium Btu Gas, in situ gasification and liquefaction, will have a target date of 2000. Liquefaction is regarded as the least promising, its main attraction being that it provides the only means of producing a liquid fuel for transport.

The technology will be described in Chapter 9. It involves a conflict between the form in which the gas should be produced. Most emphasis is placed upon SNG whereas there are those who believe that Medium Btu gas is far more suitable both for heat production and also for use as a chemical feed to the Chemical Industry.

The attraction of in-situ gasification is that it offers the possibility of removing men from the coal face, but it is relatively inefficient and expensive. However, at present it is in its infancy and very considerable research and development will need to be carried out before this method can be used upon a commercial scale.

TABLE 14 U.S. Coal Programme Funding 1977/8 Budget Outcome
1000's

Operating Expenses	1977	1978
Liquefaction	84,983	109,800
High Btu Gasification	58,704	42,800
Low Btu Gasification	39,592	47,200
Advanced Power Systems	12,800	23,800
Direct Combustion	56,826	56,900
Advanced Research and Supporting Technology	36,570	42,000
Demonstration Plants	50,600	28,700
Magnetohydrodynamics	25,989	33,000
	366,064	384,200
(In Situ Coal Gasification)	6,736	12,000

CHINA

Most reports coming out of China indicate that the main items of mining equipment will have to be imported. A recent report from the U.S. Commerce Department expressed the view that China is a probable purchaser of:

underground equipment and machinery

open pit equipment

coal preparation plants.

These needs include:

1. Face extraction systems - self advancing hydraulic roof supports - shearers - conveyors - lighting and communication equipment.
2. Increased mechanization - large trucks - blast hole drills - power shovels and drag lines.
3. Preparation plant to provide coal suitable for the steel industry.

China is a major advocate of hydraulic mining systems.

INDIA

Research and Development is in the hands of the Central Fuel Research Institute. The main effort is concentrated on four areas.

transportation of coal in concentrated form to reduce delivery costs.

conversion into secondary forms of energy and subsequent transportation as high voltage power by wire, or as High Btu Gas, liquid fuels or chemicals for industry, agriculture, etc.

use in smelting of iron and other metallic ores

conversion into domestic fuels.

India has an employment problem which limits mechanisation.

POLAND

Research and Development into underground mining is not very extensive as judged by the fact that extensive co-operation is taking place with U.S.S.R., U.S.A., the Federal Republic of Germany and the United Kingdom, all of which have large and expanding deep coal industries. Even so, research is taking place through the Maine Mise Study and Project Affect which is working towards large coal mines which permit the maximum concentration of effort and resources, with the main objective of solving water control, power supplies, haulage and environmental protection. An example of this type of large scale development is the Piast mine at Bierun with an annual output of 7 million tons.

The Jan mine, said to be completely automated, is perhaps a measure of the extent to which development has gone in Poland.

Conversion

Two areas of conversion are being examined:

solvent extraction of coal

pyrolysis

Current methods of treatment include the production of coke, town and synthesis gas and partial oxidation.

SOUTH AFRICA

There are two organizations involved in research and development.

The Chamber of Mines of South Africa, which deals with mine safety,is studying computer applications, strata control and field investigations.

The Fuel Research Institute involved with sampling and analysis of coal and coal grading for export.

In view of South Africa's increasing threat of isolation, considerable self sufficiency is being sought in the development of mining equipment just as she has developed all her other essential industries.

FEDERAL REPUBLIC OF GERMANY

The work within this section of the Federal Republic's extensive energy R&D programme is intended to improve the productivity and economy of the German coal mining industry and to lead to a higher degree of safety in mines and to make working there more attractive.

Budget

The R&D work is divided up into the following sectors, the numbers used being taken from the German programme. Reference should be made for further detail of individual projects, a summary of which are contained in appendix A at the end of this study. This list of projects has been compiled to include the name of the operation and a description of each project or in some cases a number of projects allocated to the same organisation. This should help all with a common interest.

2.1 Prospecting and Reconnaissance

2.2 Drivage Technology

2.3 Mining Technology

2.4 Logistics for Underground Operations

2.5 Coal Preparation and Surface Installation

2.6 Operational Planning Models

2.7 Mining in the Future

2.8 International Projects

A breakdown of costs for projects in hand at the end of 1976 are shown in Table 15.

TABLE 15 Mining Technology and Coal Preparation - Breakdown of Costs (DM Millions)

Sector	2.1	2.2	2.3	2.4	2.5	2.6	2.7	2.8	Total
Total Cost	26.0	33.1	76.9	18.3	42.3	11.0	8.6	6.6	222.8
Total Support Grant	13.0	16.6	38.4	9.0	22.0	8.4	7.3	1.1	115.8
Disbursements End 1976	5.0	6.3	10.2	3.1	6.4	6.1	5.8	0.3	43.2
Average Support (%)	50	50	50	49	53	75	85	17	52

Prospecting and Reconnaissance

The projects which fall within 2.1 are aimed at surveys of deposits in German coal fields and the development of exploration methods. These include speedier and safer methods but also less expensive means of identifying reserve coal fields. Reconnaissance of coal deposits included eight drillings, later analyzed in accord with standard procedures. A flameproof television camera for use in underground boreholes was developed and tested.

Projects covered by 2.2 to 2.5 include drivages, coal recovery, logistics underground, coal preparation and surface installations. These involve the development of new methods and ideas designed to meet specific situations and needs of the mining industry and prototypes for trials. Because they come into the innovation category, they are sponsored by the Ministry for Economic Affairs within the budget allocation for the Federal Government's Energy R&D Programme.

Drivage. The development technology is designed to produce greatly increased drivage rates and increased safety for the men on the job. The aim is to combine cutting and removal of rock with the vital "support work". It is considered that rates of advance could be increased and working conditions improved by mechanizing drilling, support and backfilling.

Coal Getting. The section 2.3 is aimed at adapting all the stages of coal recovery to the different conditions resulting from changes in strata and directions of coal seams. So far as coal face equipment is concerned, particular attention is being paid to the T-junction between coal face and roadway where the concentration of equipment occurs. Effort is being focused upon a reduction in the number of miners required here and the means of reducing accident hazards. This has involved three new projects being accepted for this sector.

- roads to be headed with an additional drum type shearer loader at the same height as the coal face, with haulage to be effected through a chainless drive. The effect is a smooth forward movement so reducing stress and strains normally transmitted to the coal.

- development and testing of methods of working coal deposits in inclined and steeply inclined planes. This incorporates hydromechanical mining and transport, the former employing pressure jets at a pressure of about 100 bar. The mine should already be in operation employing these techniques.

- use of extra-high pressure water jets employing pressures at about 1000 bars for loosening coal, rising to 10,000 bars for rock cutting. While dust nuisance is considerably reduced, employment of high pressures is extremely expensive in terms of energy consumption. A combination of pressure and mechanical methods appear to be much more promising. A machine designed on these lines could be used with advantage in existing coal face technology - even for thin seams.

New Logistic Systems. Such methods are aimed at dealing with an increased output of coal from the faces, as well as improved methods of man riding. Coal removal underground is achieved by belt conveyors and trains. Automated systems are being examined. Work was begun on continuous conveyors in roads with steep dips but this work has since been abandoned.

Development of computer-controlled systems for train haulage operations was begun, while pneumatic transport of dust products by pipeline underground direct to the user on the surface, was successful during trials.

Coal Preparation Equipment

This study has become necessary to meet new conditions resulting from modern mining techniques; that is higher moisture contents and the proportion of fines. These include:

- rapid measurement devices

- disposal of large quantities of froth residue in a fluidized bed.

- clarification of waste water in the treatment of ultra fine sizes, including a coagulator.

- pilot plant to reduce the quantity of middlings in favour of high quality coal, particularly where coals are difficult to produce.

Electronic data processing will receive considerable backing. This will include data regarding coal deposits and tectonics, to be stored in data banks. A system for a computer-backed processing of mines has been developed. A process computer-controlled system for the collection of technical data was proved, on-line,with particular success and has been extended to a number of collieries. The last project 2.7.1, "The Coal Mine of the Future", is intended to portray, through a series of studies, the colliery of the eighties. This will mean working at greater depths and at greater distances underground and encountering increased strata pressure and higher mine temperatures, which engineering of tomorrow will be expected to overcome.

FRANCE

Although one of the smaller producers of coal in Europe, France has mounted a fairly intensive programme through the Centre d'etudes et recherches des Charbonnages de France - Chercher.

Their recent activities can be summarized under the following headings:

- Explosives, Methane and Coal Dust Explosions
- Explosions and Fires
- Ventilation, Anti-Firedamp Measures and Mine Fires
- Noxious Dust and Pneumoconiosis
- Air and Water Pollution, problems of noise covering both under and above ground installations
- Safety Equipment

BELGIUM

This is a good example of another small coal producing country carrying out its own research and development. The Institut National Des Industries Extractives INIEX, is dependent upon the Department of Economic Affairs. Its main areas of research cover: coal, quarry industries, industrial safety and the environment. The main areas of expenditure can be seen in Table 16.

TABLE 16 INIEX Expenditure 1975-1977
Belgian Francs millions

	1975	1976	1977
General Services	47,947	51,300	61,850
Colliery Techniques	22,010	27,794	29,884
Underground Gasification	7,133	13,111	50,150
Evaluation of Quarry Products	19,444	18,444	22,857
Analysis Laboratories	4,295	11,273	12,542
Industrial Safety	9,890	12,216	14,357
Control - Atmospheric Pollution	18,425	23,446	26,060
	129,144	157,584	217,700

The large increase shown for 1977 arises from an additional 40 million Belgian Francs in respect of Underground Gasification for the first exploratory borehole at the underground experimental site.

The alternative methods are described in Chapter 9 where the Belgian activity in particular their fact finding rate is discussed in some detail. This includes a review of development in a number of countries.

UNITED KINGDOM

British research is divided into four sectors: Mining, Utilization, Scientific Control and Medical Research.

International Energy Agency

Superimposed upon this activity there is international co-operation through the International Energy Agency (IEA) in which the U.K. has been nominated to take the lead in coal research and development. The Chairman of the IEA Working Party on Coal Technology is Mr. Leslie Grainger, former Member for Science to the National Coal Board, which is the main promoter of research in Britain. This work is financed, in respect of five projects, the most important being the fluidized bed boiler project in Yorkshire, £10 million being made available in 1975 over a period of five years, with a further one million to cover other U.K. projects. Later the Grimethorpe project was increased to £13 million over a period of 7 years.

European Coal and Steel Community

The ECSC agreed in 1975 to support NCB projects through grants totalling £2.2 million, payable over three years. Updating this can be seen from the Budget and Expenditure listed in Table 17.

TABLE 17 U.K. Expenditure on R&D 1974 to 1977

	1974/5	1975/6	1976/7
Mining	4.8	12.0	16.0
Coal Utilization	2.4	4.8	5.7
Medical	0.6	0.9	1.0
Other - Scientific	0.5	1.0	1.1
Gross Total	8.3	18.7	23.8
E.C.S.C.	1.1	1.9	3.0
U.K. Effort	7.2	16.8	20.8

From Table 17 it can be seen that the ECSC contribution, though useful, is not a large percentage of total expenditure, the bulk of which is spent on mining.

Mining

This aspect of the N.C.B.'s research and development programme is carried out by the Mining Research and Development Establishment at Bretby in Leicestershire which worked on a budget of £4.6 million in 1974/5 rising to £8.7 million in 1975/6, before ECSC grants, rising yet further in 1976/7. A new method of supervision of the mining research and development programme by Development Committees was introduced under the Chairmanship of an Area Director with representatives from engineers, managers and other departments. Considerable changes to programmes were introduced and and suitable sites for trials proposed. Also a new system of funding demonstrations of proved prototype equipment from the R&D budget before being put into commercial production was successfully introduced.

The NCB's current programme has the following objectives.

Long Term. This sets sights at the year 2000 with the aim of removing men from the working districts underground by the introduction of conversion of coal "in situ", to fluid forms more easily transported and then subsequently converted into solid form. This process is particularly applicable for the extraction of coal, otherwise impossible due to thin seams, steep inclines, remoteness, depth or geological disturbances. To explore the very long term, a multi disciplinary group has been holding wide-ranging discussions with a number of research organizations in the U.K., U.S.A. and West Germany.

Medium Term. Plans here include the monitoring of recovery and automation of mechanized longwall technology. Accurate control and speed of action are seen as a means of lessening exposure of men to danger and as a result increased productivity is obtained.

Short Term. Greater recovery of reserves is seen as vital. This means reducing the amount of roof coal remaining which will increase output per machine shift. The intention is to employ more intensive layouts, re-use roadways and so reduce development. Developments to produce greater recovery of reserves using existing systems have included trials incorporating further trials of shield-type supports. The object was to obtain coal recovery at greater heights. Also, a project designed to reduce the lower limit of seam thickness when floor mounted shearers are involved. The cost of drivage is an important factor in recovery of reserves: considerable work has been taking place to speed up this process. In 1976/7 a highly successful in-seam miner was developed for work in thinner and steeper conditions; this received the Queen's Award for Technological Achievement.

One of the major factors in increased coal recovery is reliability of equipment and mining systems which form an essential step towards automation. Power loader haulage and face conveyors form an essential part. Here chainless haulage systems have been developed underground incorporating 150 H.P. modular face conveyor drives while others of 300 H.P. are at an advanced stage of development. Clutch and brake units to replace fluid couplings have also been developed.

A report on the million ton per annum coal face covering engineering and the equipment problems associated with ultra high output faces was produced during the year.

Developments. A number of projects reached the commercial stage during the period 1975-7.

- automatic control and monitoring of coal clearance conveyor systems, mechanization and integration of operations at the face and packing have been successfully mechanized. Prototypes have been proved for fully mechanized face end systems.

- proving of a shearer taking a wider (one metre) web

- 20 automatic steering installations are now working and the electric version of the in-seam miner has been proved.

- a hard rock drivage test rig designed to provide data for the design of a rock heading machine which would be cheaper and lighter than the existing types available was set up.

- successful trials at Whitwick Colliery composed of heading systems capable of disposing locally of dirt from the road without the need for transporting it to the surface, took place.

- seismic fault location systems were proved and put into practice.

- Brodsworth Colliery enjoyed success in remote controlled underground transport in the automatic operation of bunkers as well as remote monitoring of the mine environment.

- steps to reduce noise in basic design.

- a liquid nitrogen powerpack to power hand tools, with the advantage of freedom from environmental hazard.

- experiments in the use of steam for dust suppression with the conclusion that enclosure of the dust source is essential to significant improvement.

- a thickness sensor based upon natural gamma radiation from the roof strata.

- a prototype methane drainage monitor tested underground and put into commercial development.

- extensive trials on the position of the cage in the shaft enabled a system to be developed to indicate slack rope in the cage suspension chain.

- large-scale fire tests were carried out to assess the resistance of fire to various kinds of roadway linings.

- tracing the sources of underground waters using chemical and microbiological techniques.

- new instrumentation for the measurement of respirable dust and the continuous estimation and recording of ash, sulphur and moisture in coal.

- a special training centre was set up to accelerate the production of new techniques backed by ECSC money in the form of a £600,000 loan.

UNDERGROUND

Looking in rather greater detail at some of the coal face advances, the feasibility of a 1.22 metre web in a 1.22 metre seam section has been examined while a case has been made for a floor mounted thin-seam shearer for use in a 0.71 metre seam section.

Comparison trials of Dowty 4 x 300 lemniscate-type chock shield supports and Becorit caliper shields are continuing during 1978 at Markham Main Colliery, Barnsley Seam near Doncaster.

Drivage

The hydraulically powered in-seam miner is now being used. Trials have shown it to be capable of 0.23 metres advance per minute at a length of 4.27 metres between motor centres and at a seam height of 1.37 metres.

Two methods of improving roadway support using concrete are on underground trial. One consists of spraying the walls of newly formed tunnels with rapid set concrete to provide immediate support and is called "shotcreting". The other method employs large plastic bags placed behind arches into which liquid concrete is pumped to fill the space between the arches and the roadway which has just been excavated.

Underground trials of a prototype back ripping machine gave satisfactory results while in two other instances an impact hammer to break up material, and loading bucket are being introduced to gain additional experience.

Underground Transport

This important aspect of mining is divided into three parts - Mineral Transport, Man-riding and Movement of Materials.

Mineral Transport. The main work here is concentrated on conveyor transfer points to improve the transfer of coal from one conveyor to another and so move on to full automation systems. The object is to examine the effect on transfer of particle size, moisture content and dirt content of the coal.

Pneumatic coal raising shafts have been installed at two collieries, Shirebrook and Fryston, increasing considerably the capacity of the shaft.

Manriding. Pneumatically tyred free-steered vehicles are to be used in an underground trial at High Moor Colliery in North Derbyshire where both men and materials will be carried down a drift to a new part of the mine. This method requires strenthened roadway surfaces to be laid.

Materials Transport. This type differs very little from the needs of man-riding except that a linear motor will be fitted to the vehicle. This is expected to offer particular benefits since the type of vehicle does not depend for its traction on wheel adhesion, while the special bogie assists the vehicle to negotiate curves.

Elimination of dead ropes is also a goal, together with the development of special purpose rail-mounted vehicles.

COAL PREPARATION

The following development work is reported:

- Rawdon Colliery is the site of the first microprocessor system for automatic sequence control of plant operations and blending of coals. The situation with regard to plant and record of stoppages are shown on a colour VDU display.

- With the object of investigation of the benefits of automation by computer control, in coal preparation plants, a computer based monitoring system is being installed at Hall Colliery.

- The Clarometer, used to monitor the settling rate of flocculated suspensions has been developed. Flocculating reagent is added at rates which ensure optimum control and cost saving of flocculant.

- A joint venture with the U.S. Department of Energy, formerly ERDA and UKAEA (Harwell) is the evaluation of a neutron continuous sulphur monitor for use on British Coals.

- Evaluation is also proceeding, on a full scale, on two 100 tons per hour rotating probability screens for extracting dry fines at about 5 mm size.

It is hoped that this will make the washing of coal for power stations unnecessary.

- An automatic handling system for magnetite has been designed, while work has progressed on the standardization of dense medium circuits. This work includes the testing of high capacity magnetic separators.

- A new type of washbox electronic sensing and control system has been developed.

- Improvement in the reliability of deep cone thickeners has continued together with a satisfactory means of detecting the concentration of thickened solids and of controlling the discharge rate.

- Improved methods of operation of conventional filter presses for the treatment of tailings have been studied. Small scale trials, with a variable membrane plate press, indicate that drier cakes at bigger throughput may be achievable.

COMPREHENSIVE MONITORING

The main objective of the current work lies in the automation of mine operations and plant. This has been achieved recently through MINOS, a flexible and powerful modular computer system with a wide range of colliery applications. Other work of note includes:

- Three conveyor control systems now operational at collieries, with automatic bunkers forming part of the system.

- Six environmental monitoring schemes have been planned, three of these for introduction shortly.

- Programmes for mini-computers controlling or monitoring plant are now available for the control of belt conveyor transport systems. Also for the environmental monitoring of methane, airflow, gas drainage and fires as well as the monitoring of unmanned plant. Others are under development for production monitoring to locate the source and log the duration of delays and for blending and coal/stone segregation.

  Information for management control is stored on an hourly, daily and longer term basis for instant recall on a monitor screen and as a print-out.

- Success in the development of monitors for coal flow and levels in staple bunkers, is close but not yet in production.

- Underground trials for environmental systems of a vortex-type anemometer, BA4, for airflow and a high concentration methanometer, the BM2H, has now reached production stage.

- Prototypes of an instrument for colliery temperature and humidity monitoring have also been developed and reached the stage of limited production.

- The monitoring of the position of a shaft winding system in any position has been achieved at Maltby Colliery where the steel rope has been magnetically striped and cage position determined by counting the number of magnetic cycles covered with a twin magnetic detection probe.

Basic Studies.

The Mining Research and Development Establishment (MRDE) also carries out a number of basic studies covering:

Mine Environment

Strata Behaviour

Engineering Principles and Materials

Over 130 irrigated dust filters are now in use underground. A large filter with capacity of airflow rates up to 7 cubic metres per second and prototypes are now under test. One version includes the incorporation of a filter in a heading machine. A self-advancing dust filtration system has been developed for the NCB/ Dosco dintheader which has proved to be successful in a rapidly advancing drivage. Water powered dust extractors have now reached the stage of being operated on conveyor-mounted trepanners.

The use of chilled dust suppression water for reducing working temperatures is being considered now that prediction of underground temperatures can be calculated to within $\pm 1^{\circ}C$. Experiments with monolithic packing at Abernant Colliery have been successful so that roadways may now be used twice. This has followed the development of an accelerator which enables ordinary cement mix to set at rates comparable with quick setting types.

New techniques for measuring the fracture properties of steel are contributing considerably to specification and use of mining steels, while it is proposed to set up full-face trials of a powered support system protected against corrosion. This included analyses of mine waters from coal faces exhibiting corrosion problems.

A survey of alternative seismic equipment for underground use has been carried out in order to record better the reflected waves from signals transmitted through blocks of coal in order to determine the location of faults by seismic methods. In this connection, a holographic technique is being developed by University College, London.

Testing

MRDE has a testing site at Swadlincote, where comprehensive approved testing to establish the safety, efficiency and reliability of equipment and systems is carried out.

Testing projects included:

- an extensive two year programme for the assessment of heavy-duty conveyors, a co-operative effort funded by ECSC - already half completed.

- roof supports with particular emphasis on caliper shield supports plus pack hole and buttress supports for use at the face end.

- four complete face end systems set up on the surface.

- much hydraulic equipment used underground converted to run on fire resistant hydraulic fluids of the inert emulsion type. Testing has been concentrated on equipment for use with fluids containing a minimum of 90% water.

- 300 A roadway cable couplers simulating conditions underground.

- electric motors, of which 19 out of 33 failed to meet specification needs.

- non-metallic materials including conveyor belting, fire resistant hydraulic fluids, ventilation ducting and safety helmets. Test methods for fire-resistant greases are also being developed.

Training

A number of courses have been set up covering:

a) Automated steering of fixed drum shearers.

b) Cable handling.

c) Centralised underground hydraulic power supply.

d) Coal preparation.

e) Computer-controlled coal clearance.

f) Noise control.

g) Shield supports.

International Collaboration

Committee of the Mining Research and Development Establishment has representatives on corresponding committees of St BV and Charbonages de France. German and French representatives now sit on NCB Major Development Committees.

A project to improve the efficiency and safety of face end operations is underway in conjunction with Saarbergwerke A.G., while an exchange of information has been set up between Silverdale and Ensdorf Collieries.

Useful exchanges in ideas took place with the Association of West European Coal Producers - CEPCEO.

United States

Collaboration with the U.S. Bureau of Mines (USBM) developed into a 10 point programme covering;

- development and testing of a high capacity face conveyor.

- potential in U.S. for U.K. chainless haulage systems.

- exchange of respirable dust control items for testing.

- exchange of sensors for coal face interface detection.

- design of a thin seam-in-seam miner for U.S. use.

- engineering the installation of USBM water jets on a road-header.

- tests in U.S. of a USBM bendable belt conveyor.

- witnessing home tests of each others alternative power sources for underground vehicles.

- advice on a surface test facility for USBM.

- USBM engineer to visit U.K. to witness NCB subsidence engineering technology

U.S.S.R.

Exchange of visits between working parties has taken place, discussing dust control both in mines and on rock heading machines. These took place both in U.S.S.R. and in a return visit to Bretby on the reliability of machines.

Poland

The research organisation KOMAG visited Britain adding reliability of mining machinery and equipment to four fields for co-operation already agreed upon.

COAL RESEARCH ESTABLISHMENT

This research unit is at Cheltenham in Gloucestershire, being the NCB's Coal Utilisation Research arm.

Together with the work carried out at Leatherhead, the following fields of research are covered:

Coal refining - Coal derivatives - Domestic Appliance Development - Fuel and Appliance Testing - Industrial Appliances - Fluidised Combustion - Gasification - Smokeless Fuels - Metallurgical Coke - Environmental Control - Mineral Products - Scientific Services.

A breaking down of the general headings reveals the depth of the work being carried out.

Coal Refining

This covers Electrode Coke, Hydrocarbon Oils, Extraction of coal with gases, Pyrolysis and Thermal Cracking.

Coal Derivatives

The main products being produced from coal or coal spoil include: Foams, Hyload, Fire retardants, Solvents, Tars and Pitches, Activated Carbon, Resins and Heat Shrinkable Plastics.

Domestic Appliance Development

Here two areas have been examined: smoke-reducing open room heaters and use of appliances - in summer.

The field of domestic appliances is becoming all important as more smokeless zones are introduced.

Fuel and Appliance Testing

Six items make up the programme through NCB Marketing Department: Fuel testing, Appliances for domestic solid fuel, Overall efficiency of appliances, Feature fires, Chimneys and settings and a firelighter.

Industrial Appliances

Five fields of activity are listed: fluidised bed stokers, the escom high intensity stoker, underfeed stokers with automatic de-ashing, multi-shell boilers, coal and ash handling.

Fluidised Combustion

Corrosion is the main activity particularly in gas turbines.

Gasification

The gasification of coal in air or a mixture of air and steam in a fluidised bed is being investigated to produce a low BTu gas for firing gas turbines.

Environmental Control

Here leakage at lids, doors and ascension pipes in coke ovens is being checked. Also an investigation into the mechanism of nitric oxides formation during combustion of solid fuels.

Mineral Products

This sector involves what is virtually the beginning of a new industry - production of hard stone from mine stone waste. Stabilisation of colliery spoil, using cement is now well advanced. Also artificial aggregates have been produced by sintering pellets produced from cyclone ash. These contain 5 to 6% carbon.

Leatherhead Laboratories

Since 1971 the Leatherhead Laboratories have been engaged in contract research mainly on the gasification of coal and fluidised combustion.

Gasification of Coal

The main object here is to provide the necessary background information to enable the design for the synthesis gas producing part of the COGAS process to provide substitute natural gas and oil from coal.

The plant at Leatherhead uses 2 ton per hour of char to provide a synthesis gas with a low nitrogen content. No oxygen is used.

This is the process chosen by the Illinois Coal Gasification Group financed by ERDA in the 1978 budget allocation - now taken over by the U.S. Department of Energy.

When put into commercial application, the first stage would be pyrolysis followed by the gas stage and finally a methane synthesis stage.

Fluidised Combustion

Here the object of the research is to obtain necessary data for the design of power generating and steam raising plant fired by these systems. The bulk of the research will be carried out using coal but the development of residual oil burning systems is proceeding in conjunction with B.P.

The work has been supported by a consortium consisting of the American Electric Power Company, Babcock & Wilcox and Stal Loval, plus funds from the U.S. Department of Energy.

The principal test facility is a pressurised combustor which has been operated to provide information to enable a power plant with a much higher efficiency and lower emission levels of pollutants than is the case with conventional systems, to be developed. Thermal input at 6 bars pressure is 6 MW.

Other experiments involving pressurised and non-pressurised combustion are also being carried out. The thermal input is about 0.1 MW operated at 6 atmospheres. Some of the work is sponsored through the National Research Development Corporation and Combustion Systems Limited.

Oil firing as well as coal firing experiments in the fire-tube fluidised bed boiler have continued. This has a thermal input of 3.5 MW: it has been used to supply the gasification programme.

Among the consultancy work is the 85 MW pressurised fluidised combustion test facility at Grimethorpe sponsored by the International Energy Agency.

Article 55 of the E.E.C. ECSC Treaty concerning the establishment of the European Coal and Steel Community specifies that the Commission shall foster technical and economic research relating to the production of coal and the promotion of coal use and to safety in coal mines. This obligation demands that the Commission shall organise co-operation between existing research establishments.

Coal Research Committee

This consists of representatives of coal producers, research establishments, universities and trades unions who co-operate with the Coal Directorate in Brussels in choosing from a list, compiled continuously since 1967 of mid-term aid programmes for coal research.

THE PROGRAMME

Remote control technology and automation have always been recognised as immensely important and capable of fairly rapid development over a wide field. Remote controlled technology and automation for improved safety in coal mines, as well as improved performance of the coal industry within the Community, are seen as vital.

Special priority for sponsorship under the 1975-80 programme are covered by three main headings:

- operation underground

- operational management and planning

- product improvement

- new products and processes from coal

Funding

A greater demand for Community funding than exists within the pool is often sought. Then, national or international finance must be used.

Some 10% of the total research budget - over 4 m.u.a. - is to be spent on automation and remote control technology. Additional funds are available for drivage and coal recovery.

Distribution of funds is shown in Table 18.

TABLE 18 Community funds 1973 to 1976 for Automation and Remote Control

1,000's u.a.

Country	1973	1974	1975	1976	Total	%
Belgium	-	247	-	-	247	6
Britain	234	160	2,411	440	3,245	75
France	-	-	-	138	138	3
Germany	136	-	543	-	679	16
	370	407	2,954	578	4,309	100

It can be seen clearly in Table 18 that Britain enjoys the lion's share, at 75%, of total grants.

Available funds are divided among four sectors of activity. These are set out in Table 19.

TABLE 19 Division of Funds within sectors

1,000's u.a.	Belgium	Britain	France	Germany	Total
Control & automation in coal recovery	-	1,906	-	259	2,165
Automated transport	-	843	-	420	1,263
Radio underground	247	160	137	-	544
Coal preparation	-	337	-	-	337
	247	3,246	137	679	4,309

Table 19 illustrates the high priority given to control and automation in coal recovery, followed by automated transport. However, this is not a complete picture of all the effort being made in this field as has already been indicated.

This Table also indicates the way in which the four countries involved have divided the work, but with Britain operating across the board.

It should be remembered that in the first place, the funds come from the levy on West European coal and steel producers and are therefore unlikely to increase.

TRANSPOSABILITY OF RESULTS

One benefit from Community activity is the interchange of ideas resulting from co-operation between research organisations. This should ensure faster progress, quite apart from the fact that the results of the R & D are available to member countries. The INIEX - Delogne radio communication developed at the initiative of the Belgian research establishment is known world-wide. In addition to operational and prototype installations in all Community countries experimental work is being carried out in the United States and Canada, since it is particularly applicable to long range radio communication in coal mines. The wireless inductive control system developed by Bergbau-Forschung used mainly for wireless control of material transport systems has evolved from the INIEX concept and is again being taken up by The States.

A British system, developed only after very considerable effort, to monitor top coal on shearer faces is well known in many countries, while Cherchar (France) have produced a device for monitoring the coal face advance. This equipment is used very widely.

Health and Safety

Harmonisation of official Safety Regulations would do much to make the production of equipment more standard, particularly apparatus for measurement and recording: also limit switches for safety purposes. A start could be made in major areas such as firedamp and dust suppression following reasonable agreement over flameproof equipment.

Future Objectives

The main objectives must clearly be to:

- keep machines employed more continuously
- improve working conditions
- improve safety
- central monitoring of total operations

Central monitoring includes extensive pit control monitoring networks supported by additional automatic data recording and processing systems - a guide to overall management of mining operations. This will cover the monitoring of limit values, identification of trends and automatic calibration and checking.

Parallel with the phased introduction of automation must be automation and process control of surface operations which appears likely to be achieved before applications underground are completed.

Whatever may have been achieved by way of improved automation, monitoring and coal recovery, computers are able only to respond to the instructions fed into them by human beings. We need look no further than the disasters involving giant oil tankers to realise that machines are not totally reliable, while no human has been found to be infallible. Considerable research still needs to be done. Some men will probably continue to be needed underground for as long as coal has to be won. This will include generations as yet unborn and probably those who come after them.

Chapter 5

PROSPECTING AND SURVEYING – IN PERSPECTIVE

The world's coal resources have been estimated. The figures are listed in chapter 3. Some of these reserves have been classified as economically recoverable. The remainder are said to be measured.

The main object of prospecting is to make the best use of known reserves of coal: it is this aspect which will be examined here. However, some reference should be made to the need long term for adding to the known reserves. To some extent this can be forecast from a knowledge of the geological history of those parts of the world likely to have passed through the processes described in chapter 2.

However, modern techniques now make confirmation rather more positive than might have been the case a few years ago. In July 1976 Reading University was the site of the first United Nations training seminar in Remote Sensing to be held in Britain. It was the sixth in a series planned by the United Nations on space application in different regions of the world. The main object was to help developing countries use one of the advances in space technology - Remote Sensing.

Remote Sensing involves the detection by sensors in spacecraft, rockets or aircraft of reflected or emitted radiation from the earth's surface and subsequent analysis of these signals.

It is an established fact that different surfaces with varying properties emit different amounts of radiation. The study of these differences enables scientists to add to the information available on the natural resources of countries where existing knowledge is insufficient. The applications may be extremely diverse covering crops, forests, soils, water and mineral resources. It is the mineral resources which are of particular interest to those seeking additional coal reserv- es.

But this type of exploration is non productive and known as such. It has never formed more than a very small part of total coal exploration.

Here we are concerned with mining and coal production. Non productive methods of themselves produce not one ton of coal: our immediate concern is the winning of coal.

First some of the fundamentals need to be examined - methods employed in mining - exploration objectives - the role of the geologist.

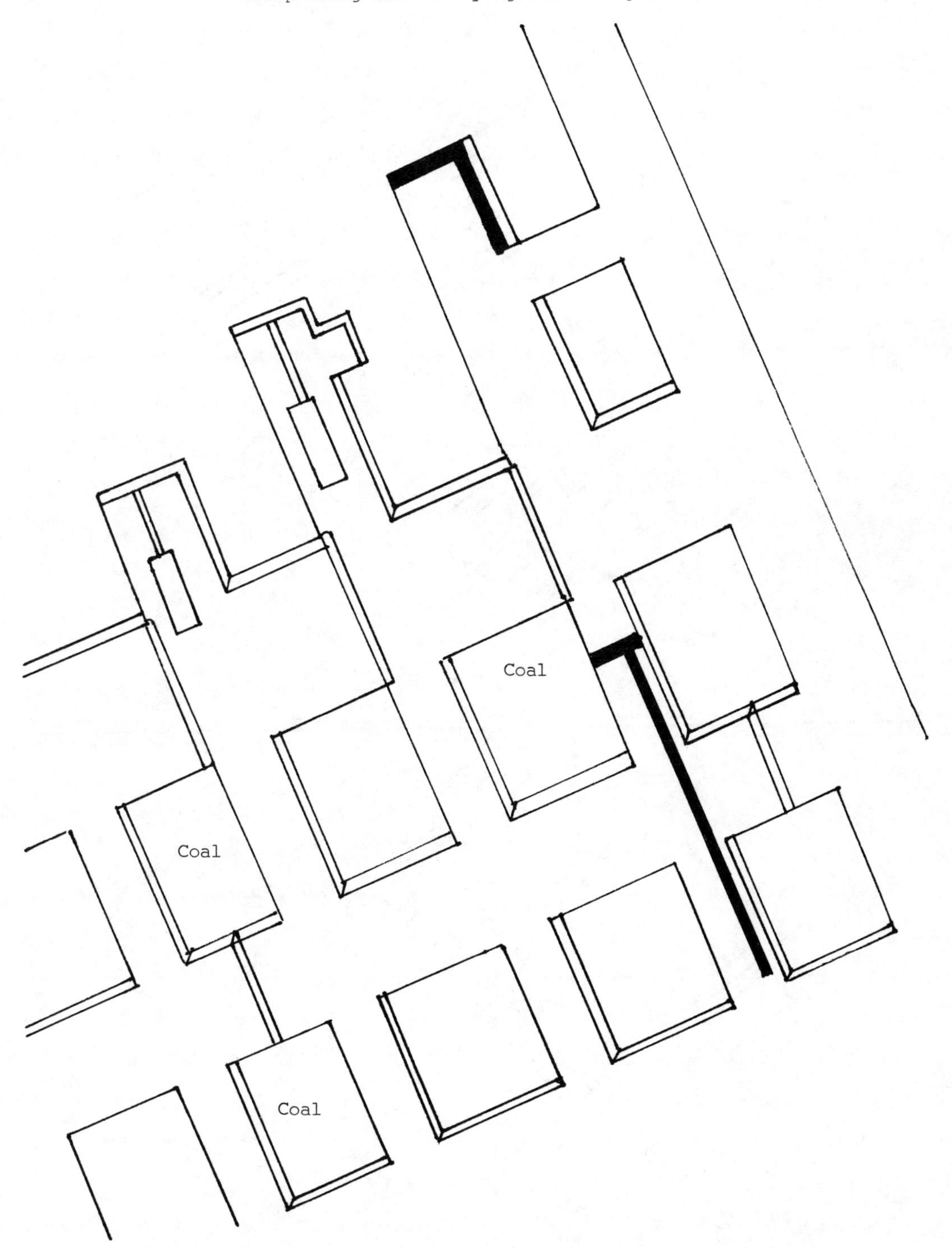

Fig. 10 Conventional room and pillar mining layout

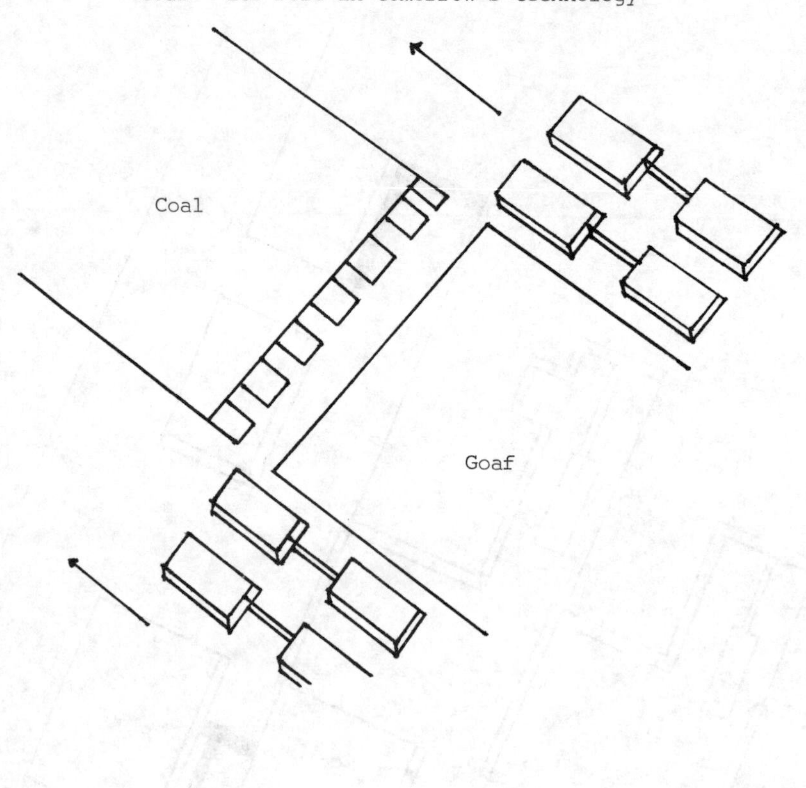

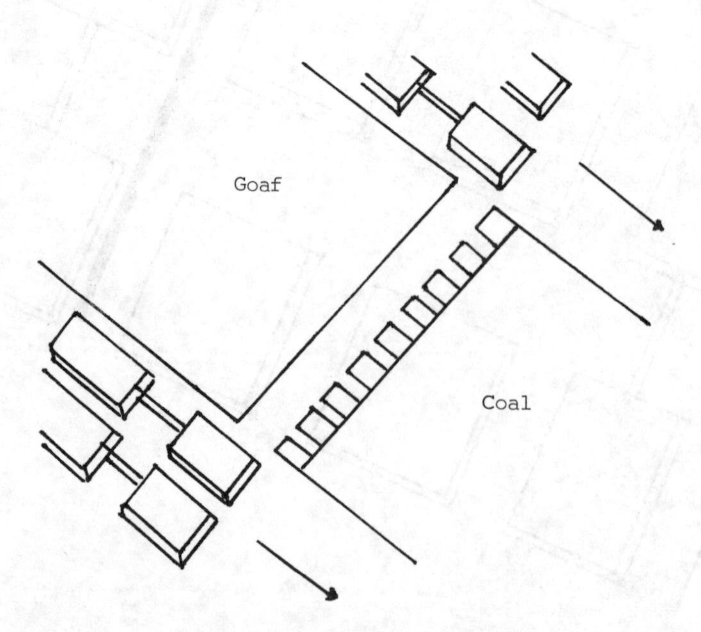

Fig. 11 Longwall mining situations

Methods of Mining

There are a number of methods of mining used world-wide. Reference has been made to most of them already, but a brief description might be helpful.

In the early days when a shaft had been sunk, coal from the best seam was mined out at the time by bord and pillar methods, returning to the shaft by a process of pillar extraction.

In the United States, the conventional room and pillar approach is still used to a considerable extent. This method is illustrated in Fig. 10. But it is slowly giving way to Longwall methods, which can take two forms.

In Europe the advance method of mining is often employed. Here the roads are driven forward with the coal face. This method is illustrated in Fig. 11. An alternative version is the retreat method of mining. The difficulties of maintaining the rate of advance of the access roads to the face have resulted in many modern mines to drive them completely beforehand.

This method of'retreat mining' is used on a large scale in both Poland and U.S.S.R. In Britain, where it has been introduced high productivity and consequently profitable mining has resulted. Finance charges are high since considerable investment is needed and much longer time elapsing before coal can be produced.

As can be seen from Fig. 12 two access roads must be driven for the full length of the panel of coal to be extracted before work can start on the face. However, in the process the long drivages provide very considerable information about conditions which will be met during recovery of the coal.

This is possibly the greatest advantage from retreat mining. Space at the face of the tunnel drivage is limited, which in turn limits the amount of equipment and men able to work there. Drivages under any conditions of manning are therefore slow.

Coal Exploration Objectives

Traditionally, exploration has taken place very close to base. The object has been to pinpoint sufficient seams which could be mined, during shaft sinking, by bord and pillar methods, returning to the shaft by a process of pillar extraction. But as winding systems become mechanised so an increase in working places was necessary to keep the shaft fully employed.

Previously, equipment was easily moved which meant that transfer from one area of coal to another was relatively simple. It was difficult to be short of coal under these conditions: the shaft was kept fully occupied.

This happy situation soon changed. The shutting down of one seam and opening of another involves very considerable physical effort in moving all the equipment from the first coal face to the second. The main purpose of coal exploration in general must be directed towards making coal cheaper to mine as opposed to exploring for more coal. Today most mines have a very long life and a great deal is already known as to the presence of coal likely to be recovered during the lifetime of even the newest recruit.

Certain priorities need to be set:

- the most accessible seams must be recovered first.

- the main access roads in a new mine must be driven towards the area which will offer the easiest coal recovery.

The changed conditions require geologists with a knowledge of coal mining rather than the traditional coal fields. The time involved today during which it is necessary to keep the pit working profitably is a multiple of that acceptable in the days of the handgot longwall era or even the bord and pillar mining. Then, the productive methods of exploration elsewhere ensured that the main access roads in a new mine were driven towards the area of easiest mining. The obstacles to good mining - the faults, the cleats and the dips showed a regional pattern which could, to some extent, have been predicted. Once winding and pumping were mechanised, then only the cleat retained its influence on bord and pillar layouts.

The old methods of mining which retained a flexibility due to short runs and high percentage extractions made the chances of correct decision making in respect of roadways in disturbed coal fields easier than would be the case in today's atmosphere of capital intensive mechanisation. While there may be considerable merit in working three or four faces in a concentrated area, a very real problem arises when a common disturbance is met. Production grinds to a halt. But disposal also brings its problems; there is a higher exploration content.

Spare capacity is therefore needed as an insurance against failure, continuity of production and an ability to match market needs. The first step is to ensure adequate stocks at the surface. Next to decide upon the number of spare face shifts carried and spread over the faces.

The Role of the Geologist

Clearly mechanisation has had a considerable effect upon the way in which the modern mining geologist thinks and the effects upon him resulting from mechanisation. They may be summarised in this way:

- the considerable increase in time necessary for exploration so that disturbances may be detected sufficiently early and hold ups in production avoided.

- built-in insurance must include the stocks of coal carried, as well as additional people from spare coal face capacity to spare face shifts.

- concentration which results in loss of flexibility in change of plan or direction because of the capital intensity of the nature of the modern operation. Also a greater dependence upon local knowledge of geological disturbance. This applies particularly to "prop free front" support systems.

- the change from the mining engineer with a broad knowledge across the board to specialists covering much narrower fields such as:

  electrical - mechanical - strata control - preparation - safety - ventilation and general planning.

As the chief geologist to the NCB put it: "The old style geologist has had to become a specialist coalmining engineering geologist, who now has to use the acquisition of his geological data, not as an end in itself but from a cost/benefit mining engineering standpoint. He has had to reconsider what is implied by coal reserves in a multi-fuel energy economy and to rethink the nature, place and function of exploration." He should know.

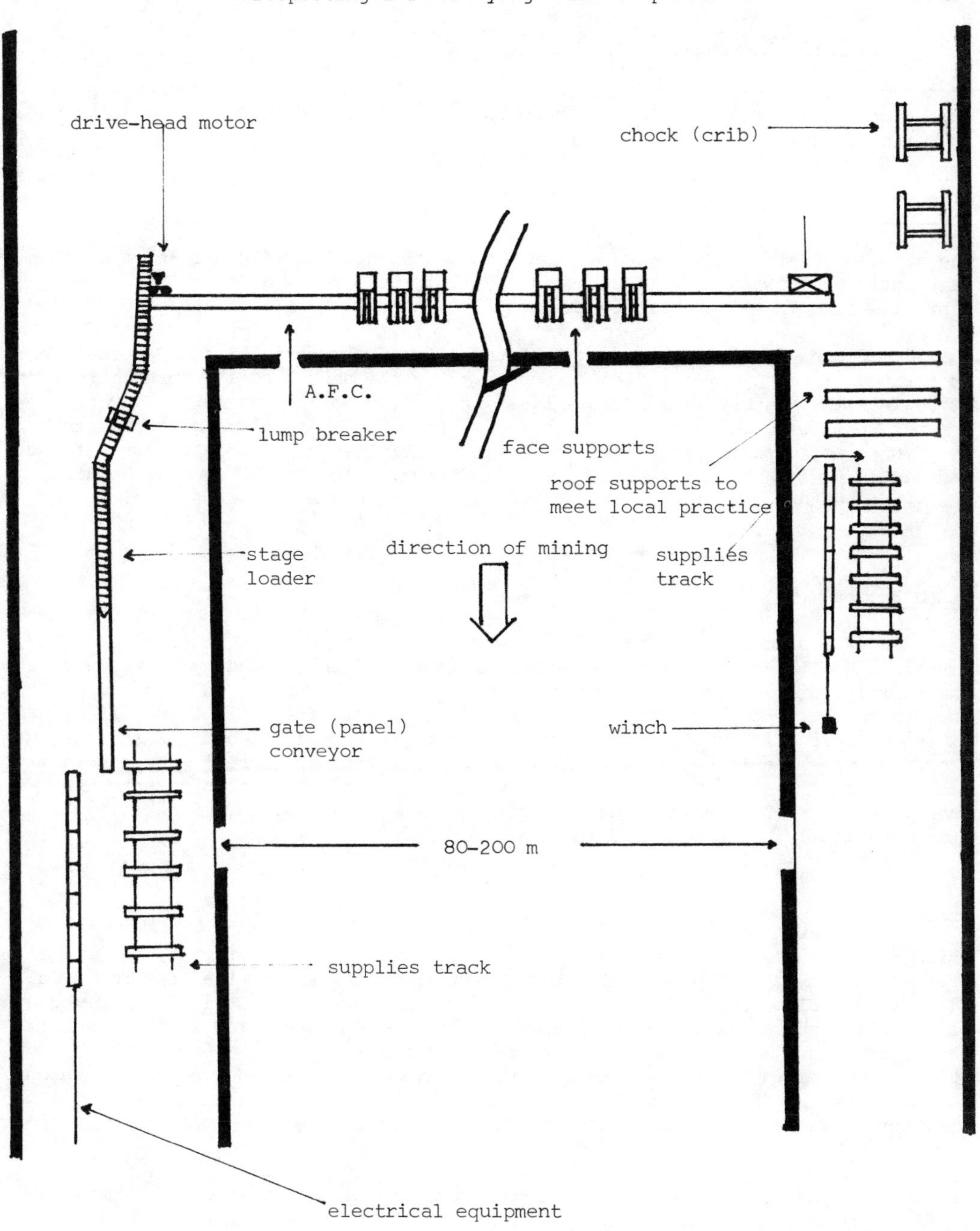

Fig. 12 Plan of retreat mining and associated equipment

MODERN EXPLORATION PROGRAMMES

There are clearly a number of factors which must be taken into consideration when such a programme is undertaken. These include coal reserves, present output and market prospects.

Coal Reserves

The assessment of reserves are necessary to decide for how long present operations may continue and when it will be necessary to open up a new area. This assessment will anticipate the move to the next location.

But it would be an error to assume that the life of a particular area or coal face is arrived at simply by dividing the reserves by the annual rate of production. To begin with, reserves are not identical in quantity or dimension. They represent coal to be recovered over a broad spread of conditions ranging from thick to thin seams, good to poor roof and floor conditions, undisturbed seams or those containing faults which take on a number of forms. The conditions may be deep, dirty and sulphur containing, making them just within the limits of economic recovery.

Human Factor

The life of a seam or mine will clearly depend upon the rate of extraction. In turn this will be a reflection of management and the general energy background which exists at the time. If the climate is one in which oil, gas and nuclear energy are cheap and abundant, coal productivity will suffer. However, if oil is expensive, gas rising in price and nuclear energy at the crossroads of indecision, then coal may come into its own, provided the incentive is there. This normally takes the form of payment for coal recovered and it is interesting to note that when the British Miners insisted on part payment by this method, productivity rose immediately. In mid 1978 there was a surplus of coal.

Thin Seams

While the revenue per ton of coal recovered is proportional to the volume of coal worked, the cost per ton of coal mined bears a relationship to the area involved. Clearly the volume depends upon the thickness of the seams. It is in fact the variation in thickness on the underside of the average seam which is the governing factor in productivity, however thick the maximum may be. Conversely, the height which can be mined is governed by the amount by which it exceeds the mine opening permitted by the depth and rock strength of the surrounding strata. This will be very considerable as compared with the contribution of even the most sturdy support systems.

As was explained in chapter 3 the minimum thickness of a seam which will qualify it as workable or economic is governed by the size of machines and the ability of men to mine it.

Seam thickness or variation in thickness bear little relationship to disturbances originating from the formation of the original strata. Further increase in thickness above that which is economically workable will have little effect upon productivity by comparison with the other factors which have been discussed.

Some of the additional factors will now be considered.

Fault Disturbances

Capital intense mining methods demand greatly increased minimum runs, where longwall systems are involved. It is, therefore, more realistic to talk about the cost per ton covering a yearly basis than for any shorter time scale. This illustrates the impact which unexpected faults will make during the life of a mine - measured in terms of annual production. It is the experience in Britain, other than in South Wales, that the faults in seams following the first seam worked can be anticipated. This makes it possible for alternative production panels to be made available to replace one another, continuous mining being achieved without the heavy insurance mentioned earlier or disturbance costs being necessary. Even so it is the experience in Britain that productivity in a given year falls as the incidence of faults rises above a certain threshold value, somewhere around four faults, of more than ten feet throw, per square kilometre. Ironically no benefit appears to be derived from thin seams with a lower than threshold incident rate.

In general terms, the production rates appear to be inversely proportional to the intensity of faulting. There can of course be other factors such as sedimentary disturbances which will cause problems.

Abandonment of Thinner Seams

Apart from mechanisation the only increase in productivity achieved in Britain in areas where seams were thinner, was by abandonning them for thicker seams and concentrate work there. But the time must come when the only way in which productivity can be sustained is to open new mines.

This is one of the problems which technological research and development must solve by changing mining methods, so as to make the move to new mines unnecessary.

Improvement of Face Capacity

When a colliery is experiencing geological problems its ability to produce coal will depend upon the number of faces available to provide the planned level of daily output. Bearing in mind the high level of face output which modern mechanisation makes possible, it is clear that idle machines and idle miners will have a marked impact upon colliery profitability.

A detailed study as to causes of delay at the coal face was carried out in Britain between 1975 and 1977. Some 1,200 machine shifts of delay study data were examined. It was found that there were ten major causes of face delay. These have been listed in Table 20.

The possibility of re-using gateroads to serve two adjacent longwall faces is being investigated due to the considerable potential benefits. These include:

- reduced drivage development
- less risk of spontaneous combustion
- increased extraction of reserves

One key factor concerns the number of strata abutment pressure zones which will occur. This should be limited to two only, behind an advancing face but abandoned after the second face has been worked. Attempts upon this are likely to be counter productive.

TABLE 20 Top-ten Delays in Face Operations

Order of Significance	Reason for Delay	Average Frequency	Average time/ occurrence (minutes)	Number of Areas where this delay is in the Area "top ten" table
1	Advance/set supports in bad ground	1 in 5 shifts	20	6
2	Maingate stable/ entry not ready	1 in 8 shifts	26	6
3	Machine cable damaged/replace	1 in 35 shifts	100	7
4	Trunk belts - running faults, slipping, sticking, fast, out of line, etc.	1 in 2 shifts	6	5
5	AFC chain fast	1 in 9 shifts	24	4
6	Late start - excess travel and prep.	1 in 3 shifts	8	4
7	Trunk belts - broken/repair	1 in 15 shifts	33	4
8	Trunk belts stopped - unknown reasons	1 in 1 shift	2	2
9	Early finish	1 in 6 shifts	12	5
10	AFC chain broken/ damaged	1 in 32 shifts	63	2

From Table 20, it can be seen that five of the reasons for delay are associated with the movement of run-of-mine coal involving AFC and trunk belt problems. Three of the delays are due to management or organisational problems. These include delay in the maingate stable, late starting and early finishing. Clearly starting and stopping of shifts on individual coal faces can result from the distance which must be travelled to and from the face. Two delays were due to operating difficulties - machine cable damage and roof supports in adverse conditions.

From this it was necessary to seek the answer to four questions:

- what losses do they represent?

- how often do they occur?

- how much time is lost?

- what is the incidence in other Areas as a prime cause of delay?

Losses

It was found that the ten categories represented an average daily delay of 26 minutes per shift or 22% out of a total lost time of 116 minutes per shift. In terms of lost production each of the ten classes of delay were estimated as causing losses of about one million tons of useable coal annually.

Frequency

The average frequency was found to vary from once per shift to one in thirty five shifts. Cable damage and chain damage however present a lower incidence - about once in every three to four weeks of operation. Three delays involving trunk belt running faults, unknown stops and late starts are relatively common, about once per day or once per shift. The other five categories occurred less frequently, one each between two and five days.

Time Lost

This appeared to vary from less than a minute to spanning a number of shifts. The incidents within that range can again be seen from Table

Incidence in other Areas

The top four categories and the ninth - early finish - appear to be common to a large number of Areas.

The report highlighted clearly five fields of activity:

- support of the roof
- AFC and trunk belt maintenance and operation
- care with machine cables
- organisation and control of stable/ entry operations
- monitoring late starts and early finishes

Number of Coal Faces

The number of faces is only one of the factors which needs to be taken into consideration when attempts are made to improve face capacity. The number of shifts worked by machines and the number of facemen, as well as the balance between face shifts, has a considerable influence. The method of mining also makes its impact; the amount of retreat working and seam section being extracted also increases capacity.

British experience over six years from 1970 to 1976 showed a loss of 85 longwall faces from 807 to 722, while daily output per face (DOF) fell from 613 tons to around 575 tons. However, by November 1976 a figure of 613 tons was reached once again.

It must be remembered that 1974 saw the miners strike, after which the number of faces rose once again to 750. This increase of 28 was equivalent to a yield of 3¼ million tons, at an annual rate, showing the need towards increased development work at collieries which have shortage in face capacity.

It is also interesting to note that the number of faces in the equipped but temporarily stopped classification rose from 43 in November 1975 to 54 in November 1976, equivalent to some five fully equipped faces in each Area not working.

It was clear that the monitoring of failure to comply with planned face changeover programmes is essential, particularly:

- reasons for faces finishing earlier than planned
- movements in starting dates for new and recommissioned faces. Also the causes for any changes as well as the effect with respect to Area colliery weekly outputs

Attention should then be focussed on the action required to bring back the number of faces to that originally planned.

Flexibility

Provision of the right number of coal faces to give flexibility and choice in face room is absolutely vital. This may involve the development of spare faces and longer replacement faces. It could involve a larger number of faces worked on fewer shifts each day to improve flexibility.

The benefits of increased flexibility are obvious, namely; overall daily and weekly outputs.

Spare Capacity

The provision of flexibility in spare face capacity makes it possible for there to be a speedy response to geological changes. These spare capacities may include the use of "face on" or back of the arches methods of coal face development: by selection of minimum face heading widths and the use of profile types of face heading machines whether or not armoured flexible conveyors and powered supports were included too.

Modern coal faces can be designed to stand for very long periods, including those caused by labour disputes.

Salvaging of Coal Faces

This important activity is complementary to ensuring that workable seams, sufficient faces and men are available. Taking a figure of 500 faces, the number which cease to be producing units in a normal year, there are probably some 48,000 powered roof supports employed, quite apart from general coal face equipment, roadway materials and services.

Teams must be set up which are retained exclusively on salvage until the job is completed. As a result individual members of the team become expert in salvage. It may be significant that their rate of absenteeism is much lower than the average level.

The withdrawal of items of equipment and materials from a coal face and its immediate roadways is a necessary part of the process of replacement new faces.

These salvage operations must be planned by a team, which should include the manager, mechanical and electrical engineers, surveyor, safety engineer and any other relevant specialists. The working party should examine the face and district services and then draw up a plan to cover the withdrawal back to the sealing off position. Their plan will need to include:

- the final position of the face taking local conditions into consideration
- the list of face and gate equipment and materials to be salvaged, with details as to their disposal
- the method of operation, particularly support rules for all phases
- the face and gate transport systems and transport rules
- manpower and training requirements
- salvage equipment required
- environmental problems including ventilation, spontaneous combustion, methane drainage and dust control
- supervision responsibilities

Rules and instructions must be properly drawn up and issued to ensure the safety of all involved.

However, the plan is one thing, the execution, another.

What is really important is that the conditions which will actually exist when production has stopped are clearly appreciated. The main points which experience has shown to be vital are:

- the face and salvage support systems decided upon, must be possible in the conditions to be met at the time.
- transport arrangements must be adequate.
- salvage teams must be properly trained.
- support and transport rules must be finalised and issued.

While it may seem to be unnecessary, it should be confirmed that the equipment to be recovered is accessible and available.

In this connection, the preparation work carried out as production is about to be phased out will affect considerably both the safety and smooth running of the salvage operation. Experience has shown the following measures to be particularly applicable:

- the face line should be kept as straight as possible.
- the maximum height possible should be attained by cutting above the usual extraction height over the final four metres of the advance.
- additional support must be introduced for poor roof conditions and cater

for goaf flushing.

- rips and stalls must be advanced to assist in dismantling the power loader and subsequent loading and transporting of the face equipment.
- face end lighting and housing for the face haulage engine are necessary.
- ensure that powered roof supports are adequately maintained and tested to provide full retraction.
- power services necessary for the salvage process must be maintained.
- clear all gates of unnecessary materials and bring up the equipment necessary for the operation.
- install additional communications, telephones and signalling systems.

Progress against the programme contained in the plan must be monitored so that schedules are maintained.

SEISMIC TECHNIQUES

Clearly good management, forecasting of probable disturbances from accumulated knowledge of the area plus continuous working form the most efficient method of exploration. However, the time does come when new areas must be opened up and new mines created.

First the area believed to be worth exploiting will be surveyed by conventional methods to confirm the original indications that the area is rich in coal and worth developing, bearing in mind the very considerable capital which will have to be invested.

This will involve sinking one hundred or more boreholes a year, depending upon the amount of fresh coal required. As the drillings are completed, then a succession of cores will be brought out revealing a rich core or disappointment. But that is not the end of the exercise. Continuity of coal seams is vital, or at least knowledge of the steps which will have to be taken when faults appear.

Seismic survey techniques have been used for some years, but the view is growing fast that surveys of higher resolution are vital if errors which have clearly been shown to be expensive, are to be avoided. When a large project involving some hundreds of millions of tons of coal is at risk mistakes must be eliminated.

Equipment

Basically a 40 man crew employs up to 5 drilling rigs to place small dynamite charges 10 to 12 metres deep at intervals of 20 metres. Forty eight geophone receivers are placed at 10 metre intervals on the survey grid, usually about one square kilometre in area. As each charge is individually detonated, the seismic signals reflected from the various coal seams are recorded in a 48 track recorder. The recorder itself produces a record tape for analysis, usually on a special-purpose digital computer.

The survey is usually carried out so that half the survey lines are perpendicular to the suspected trend of faulting. The total area to be covered will vary according to the amount of information required.

In one particular instance in Britain, 250 kilometres of seismic lines covered 285 square kilometres on an open reconnaissance grid. A broad pattern of faulting was shown moving in a particular direction with an average of about three faults per square kilometre. As a result a general pattern for the field was worked out, clearly indicating that five deep mines feeding into one main haulage system, composed of twin conveyor belts, were required.

Once the general location of each mine had been proposed, intensive seismic surveys were carried out around each site. On one occasion the seismic team discovered a fault with a throw of about 50 metres. As a result, the plan was altered, which must have saved very considerable sums of money.

In another situation at a depth of about 700 metres seismic surveys revealed a major fault with a 5 metre displacement in the fields two deepest seams.

The methods employed follow very closely those used by the oil companies. With coal, the response of seismic waves by geophones is measured at 1 - millisecond intervals, whereas in the case of oil it is 4 - milliseconds. Again in the case of coal, geophone stations are spread about 10 metres apart, whereas in the case of oil the gap is 100 metres.

Even so, an increase in the number of monitoring stations is no guarantee that additional information will result. The signal frequency must also be increased, probably by reducing the size of the dynamite charges forming the sound source. Charges probably fall between 125 to 500 grams in weight against 20 kilogrammes used for oil.

As the pulse length is reduced, so resolution appears to increase. As size of charges change, so must the stationing of the equipment; groups of geophones otherwise tend to produce a blurred effect upon the data, resulting in poorer resolution.

Single stations, rather than a group, are required. In turn the scaling down in an attempt at obtaining more shallow detail, the impact of waves called ground roll and static errors become increasingly critical.

Old Mines

Apart from using seismic lines for new mines, they can also be used in exploring new reserves at existing mines. About 90 kilometres of seismic lines, at one mine, indicated 50 million tons of recoverable reserves. This survey took ninety days over five years and cost about £1,500 per line - kilometre.

Seismic aids can only indicate what lies ahead and the problems to be overcome. In the ultimate exploration really boils down to having the mining force and its equipment in the right place at the right time working continuously. Then and only then will coal be recovered at the cheapest price.

Chapter 6

MINING PLANT AND EQUIPMENT

The construction of a new mine provides an opportunity not open to those mines which have been in existence for a number of years - to begin again. In this way it is hoped that modern methods, techniques and equipment will enable those recovering the coal to enjoy the benefits which none of the others experienced save by an extension of the existing facilities.

The steps to be taken and new opportunities will be examined in this chapter, covering the main activity required when a new mine is opened. Some, but clearly not all, of the items of equipment will be available, so that production may be continuous and at a rate which brings an adequate return on the capital invested. A list of many of the suppliers of such equipment, together with their addresses, is to be found in the appendix, as well as the products which they offer to the mining industry.

Deep Shafts

For many years it has been the practice to line mine shafts, particularly in Britain. In many other parts of the world especially where few difficulties are expected it has often been the custom to sink rectangular shafts either unlined or containing timbers. However, increased depths have meant higher pressures, including the likelihood of water, so that there is now a world-wide move towards lined circular shafts.

Certain fundamentals have been listed which will do much to ensure optimum speed in shaft sinking:

- maximum possible knowledge of the ground to be excavated
- appropriate design parameters for the conditions of the area
- careful planning and timing of the whole operation

Preliminary work will consist of coring, logging and formation, evaluation and directional control.

Logging and formation evaluation. A number of specialised techniques have been applied over the years and include :

- Drill-stem tests which isolate any part of the hole and enable measurement of flow and flow pressure to be obtained for fluid bearing strata.
- Resistivity Logging which defines porous and permeable zones.
- Gamma-ray Logging which makes use of the natural gamma rays existing in the formation or introduced from a nuclear source in the logging tool.
- Neutron Logging normally run in conjunction with a gamma-ray log. The amplitude of the neutron curve would indicate the presence of oil and water.
- Temperature Logging which involves continuous recording of the temperature at each depth of core-hole.
- Caliper Logging which provides a guide to the amount of cement required when cases are grouted in.
- Formation Testing which indicates pressure data in the zone under test and is obtained from formation fluids which are recovered.
- Bond Logging which is of particular application to the sealing-off of freeze holes above and below the zones which are frozen.
- Sonic Logging measured in micro seconds records the time required for sonic waves to pass through one foot of the formation.

Directional Control. Rotary drilling is involved guided by a laser beam. This is a more recent development replacing the traditional plumb line, theodolite and optical plummet, all of which suffer disadvantages in practice.

Automatic laser plumbs are instruments which generate a vertical beam of light even if the instrument support face tilts. The main feature of auto-plums are:

- low powered He Ne laser source, with associated optics that produce a substantially parallel beam.
- detection of the beam by a visual screen so that its position may be related to a structural reference.
- automatic setting in the vertical direction achieved by means of a non-magnetic pendulum system which controls the pointing of the beam.
- self cleating capability incorporating a bearing so that the whole pendulum assembly may be rotated 180° around a nominally vertical axis so permitting the true vertical to be established.

A typical shaft section showing a four deck scaffold and mucking unit is illustrated in Fig. 13.

Diameters of shafts sunk worldwide generally cover diameters ranging from 2.74 metres to 9.6 metres with depths to 2042 metres, taking all types of mines into consideration.

Modern techniques employed, include the use of multi-deck scaffolds to permit

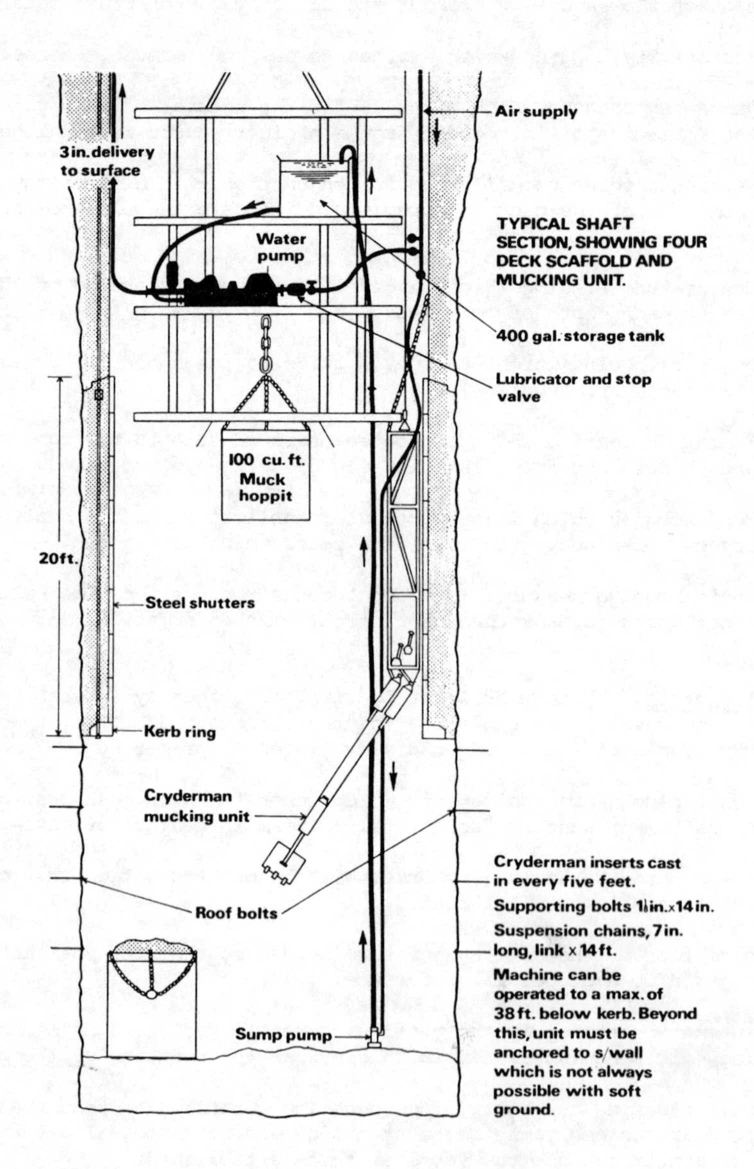

Fig. 13 Shaft Section

simultaneous sinking and walling, hydrologically operated mucking units and stabilisation of water strata by grouting.

Explosives in Coal Mining

Choice of explosives is governed by Government regulations which must pass official tests and then be classified as permitted explosives. The general properties are discussed later in chapter 8. They can be used for Shaft Sinking, Drifting, Ripping and Coal getting.

Shaft sinking. One particular precaution is demanded whenever the shaft is in close proximity to or passing through a coal seam involving gassy conditions and when methane precautions must be taken. Then "Permitted" explosives must be used.

Drifting. The same principles with regard to "Permitted" explosives apply here. Where rounds of shots are fired the maximum delay period allowed in Britain is five seconds. It is considered that this is an interval sufficient to avoid risk of a possible blown out shot fired on the later delays causing an ignition of gas or dust released by earlier shots.

Speed of advance is vital. Drilling may be either by hand held jack hammers, air leg mounted drills, post mounted drills or drills mounted on drill rigs or jumbos. The method employed will largely depend upon the size of the tunnel and the hardness of the rock. Drilling Jumbos result in a faster rate of drilling, but take longer to move during blasting and loading. It may therefore be preferable to use air leg mounted drills. Tungsten carbide tipped drills are used to obtain long life and high drilling speeds. Wet drilling will help to reduce dust particularly when a number of drills are operating simultaneously.

Explosives used will have high power, high density and high detonation velocity. Permitted explosives would include Polar Ajax type while Non-Permitted, when allowed, will involve Special Gelatine 80% Strength or Belex. Electric shot firing is normally used for drifting. It is safer and faster.

Ripping. In all mines some ripping is needed to provide extra height for roads or airways where coal has already been extracted. Since an extra free face is always present, no cut shots are needed. Here, the blasting operation is simpler than in drifting.

The process depends upon the siting of shot holes so that the roadway will be square. Siting will also govern the size of coal piece so that it can be handled easily. Figure 14 shows the correct hole placement for a number of situations where blasting is involved. A brief description is to be found above each diagram.

Coal getting. In many mines coal is first holed by a mechanical cutter, before blasting can take place, so that an additional free face is exposed, so reducing the risk of blown out shots. A reduction in dust and increase in the yield of good lump coal results.

Hard coal requires a medium strength explosive such as Polar Ajax or P3 Unigal - ICI products.

Softer coals can be readily blasted with low strength gelatines, semi gelatines

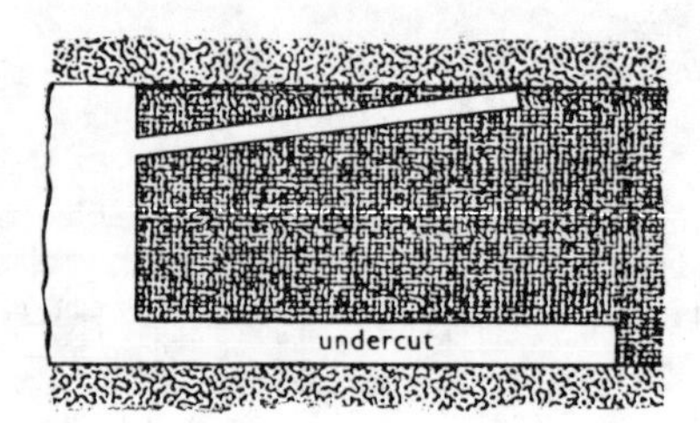

Correct position of shothole

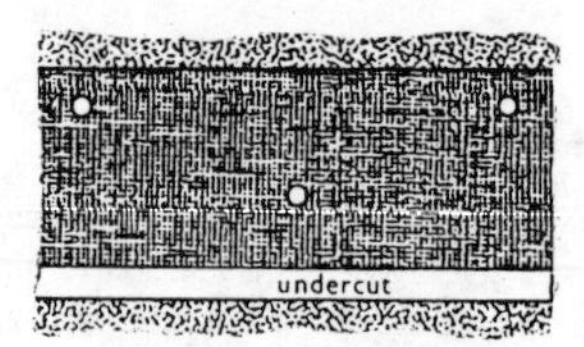

Hole placement in narrow working

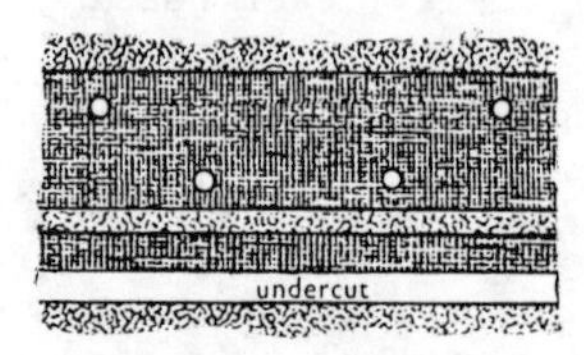

Hole placement in narrow working when dirt band is present

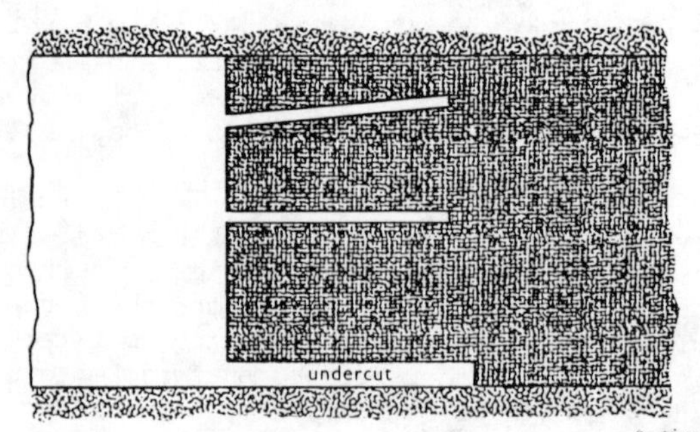

Two rows of holes are used in thick seams

Hole placement in 3.7 m (12 ft.) wide heading

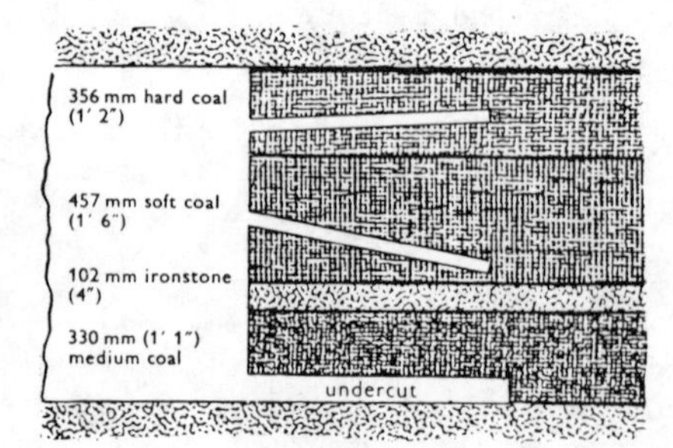

Two rows of holes may be used where a hard ironstone rib occurs

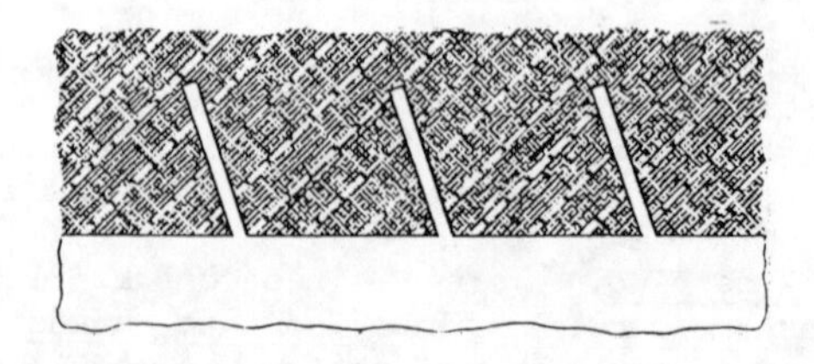

Shotholes angled to cross the cleat in a longwall face

Fig. 14 Hole placement in blasting operations

or P3 Unigex A and Unipruf. Fig. 15 illustrates a number of situations in which explosives are used in coal mining. Descriptions are provided with each sketch.

Blasting Explosives

The essential characteristic of all explosives is that on initiation they react immediately to produce large volumes of gases at high temperatures. The almost instantaneous release of these gases creates very high pressures. The reaction proceeds very rapidly, is self-sustained and is maintained throughout the mass of explosive when set off within it.

Modern explosives fall into three categories: initiating explosives, high explosives and low explosives.

Initiating explosives. This type of explosive is extremely sensitive and easy to explode. It is used to start the detonation of high explosives. When ignited the shock is intense, but local, and capable of starting the reaction in the less sensitive high explosives.

High explosives. This category detonate at velocities between 1500 and 7500 metres per second, with large volumes of gas being produced at particularly high pressures. It is the shock waves initiated by the detonator is transmitted through the coal or rock. It is followed by the high pressure gases which complete the breakage. High explosives are used in most circumstances except where a much milder reaction is needed.

Low explosives. Here the reaction involves a very rapid burning of the explosive composition but without the production of an intense shock wave. A flame or spark, normally sets them off, from a safety fuse.

Properties of explosives. The properties of explosives are important. They must be viewed in terms of strength, velocity of detonation, density, water resistance and fume characteristics.

Strength. This is a measure of the amount of energy released. The grade strength is the percentage of nitroglycerine in a straight nitroglycerine dynamite which produces the same effect as an equal weight of explosive.

Cartridge strength. As applied to an explosive it is the percentage of nitroglycerine in the straight nitroglycerine that produces the same effect as an equal volume of the explosive.

Velocity. This is the speed with which the detonation travels through a standard column of explosive. With most commercial explosives, this is 2500 metres per second.

Density. This property of an explosive depends upon the constituents from which it is made. Where a high density explosive is involved, the energy is concentrated. This is suitable for tunnelling and for use in hard ground. Low densities

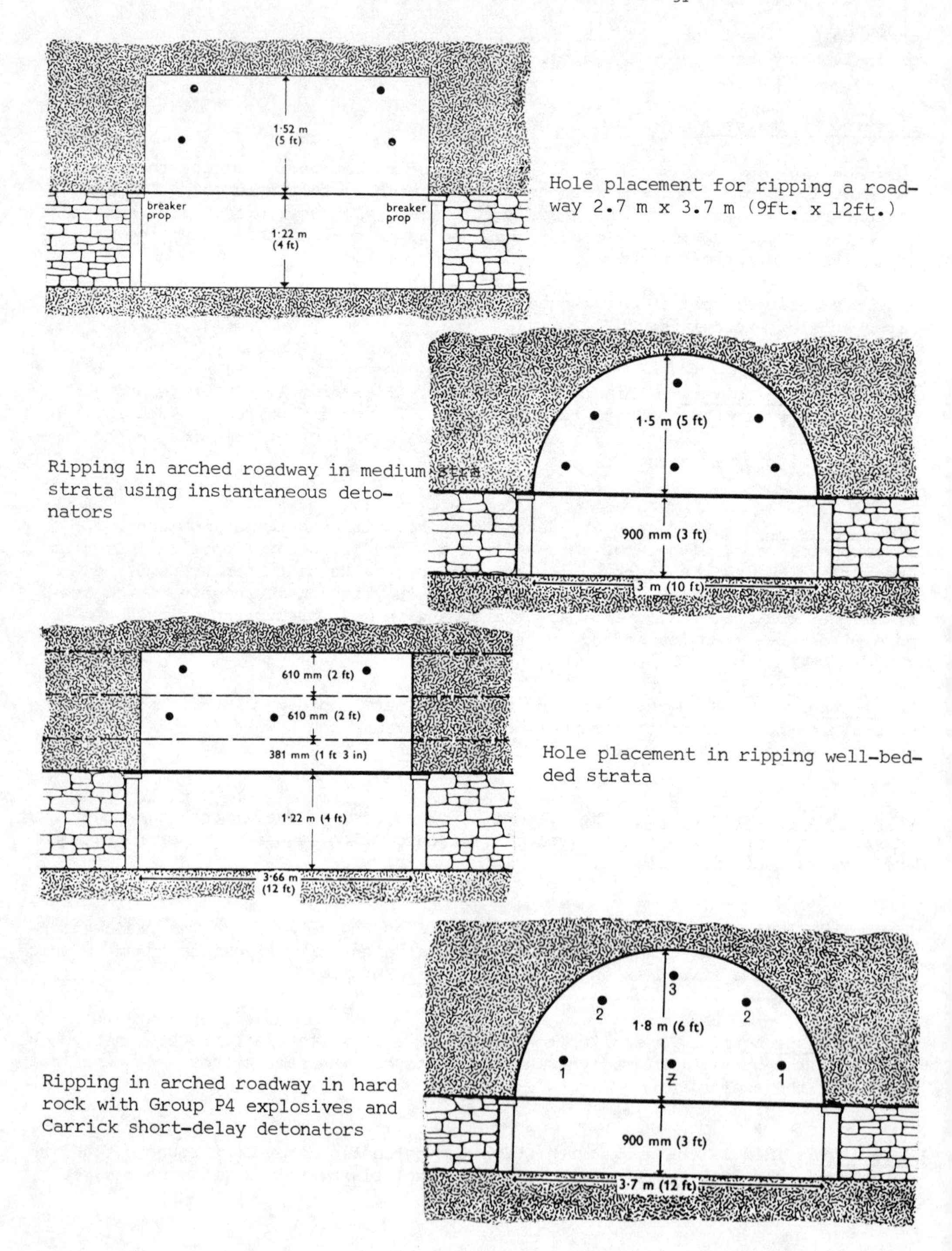

Hole placement for ripping a roadway 2.7 m x 3.7 m (9ft. x 12ft.)

Ripping in arched roadway in medium strata using instantaneous detonators

Hole placement in ripping well-bedded strata

Ripping in arched roadway in hard rock with Group P4 explosives and Carrick short-delay detonators

Fig. 15 Use of explosives in coal mining

are involved where excessive fragmentation has to be avoided.

Water resistance. This is particularly important when blasting is set to occur in areas where conditions are likely to be wet.

Explosive which has a low interest water resistance may be cartridged so that their resistance is increased.

Fume. Most explosives are adjusted to control toxic fumes but certain explosives for special purposes are not compensated for in this way. Their use should be restricted.

Explosives for coal mines. Specially designed explosives are used when gassy or dusty situations are involved to reduce risk of accidental explosion of fire damp or coal dust.

Stability. Special stability tests are carried out before they leave the factory. Careful choice of raw materials ensures stability.

Gelatinous Explosives

There are a number of types and grades of gelantinous explosives.

Blasting gelatines - high density, excellent storage unaffected by water, particularly suitable for hard rock, metal breaking and under water work as well as mud capping. Fuming is low.

Gelatines - main constituents are nitroglycerine, mitrocotton, sodium nitrate and cellulosic materials. The main characteristics are plastic consistency, high density, good water resistance, freedom from noxious fumes and good storage properties. Strengths range from 75% to 45% strengths in 5% dimensions.

Special gelatines - contain nitroglycerine in their make-up as well as nitroglycerine, nitrocotton, sodium nitrate and absorbents. The possess high density and good fuming properties, but are poor under water for long periods. Strengths range from 90%, in units of 10, down to 40%. They are vital blasting agents particularly in coal mines especially drifting and tunnelling. Fume characteristics are good.

Opencast gelignite - a special gelatine type explosive containing nitroglycerine, nitrocotton and ammonium nitrate. It is a powerful high density explosive with good water resistance particularly suited for opencast.

Ammon dynamites - contain nitroglycerine, ammonium nitrate and carbonaceous ingredients. They have low densities, relatively high cartridge counts, good fume characteristics as well as storage and water reistance properties. The cartridges are protected with molten wax. Strengths range from 60% down to 40%. They are suitable where blasting conditions are not too severe. They are produced in powder form.

Blasting agents. This group is used for overburden blasting in opencast mining. They are very safe to handle and transport due to their composition which lacks sensitivity.

Ammonium nitrate mixtures are popular in opencast mining but suffer from a number of limitations: low water resistance, low density and low bulk strength as well as low velocity of detonation.

Slurry explosives. These provide the answer to the shortcomings of blasting agents. They contain water as an essential part forming a solution which is thickened into a gel while the solid content remains in suspension, providing the necessary water resistance. Slurry explosives usually consist of a gelled acqueous solution of ammonium nitrate and sodium nitrate together with sensitisers held in suspension.

A number of accessories will be required to operate the explosives.

Underground Roadways

A tunnel as applied to coal mining is an underground roadway of not less than 6 cubic metres in area of cross section, driven in solid strata made up of coal or stone or on some occasions both will be involved.

In the Coal Mining Industry there is an increasing interest in the drivage of large cross-measure drifts for new mines or else for use in modernisation programmes. Gradients are often involved with wide variation in the type of strata to be tackled. However, with a move towards retreat mining,machines are needed to develop headings rapidly, in the same place as the seam forming the roading where the final height will be 5.5 ft. - 1.7 metres.

Clearly the aim must be to drive roadways by machine for which there is a choice of equipment.

The type to be used will depend upon

- the nature of the strata
- size and shape of roadway
- whether or not in the plane of the strata
- the gradient involved

There-are some five methods used to drive roadways.

1) Continuous and also in-seam miner - applied to coal

2) Dirtheader machines in coal and adjacent strata up to a compressive strength of 8000 psi.

3) Boom or selective heading machines of the road header type capable of dealing with rock up to a compressive strength of 12000 psi.

4) Full-face tunnelling machines which can deal with harder rocks up to 35000 psi.

5) Drilling and shot-firing - which have already been discussed.

Continuous and In-seam Miners. These machines are coal headers only. They include a number of types of Continuous Miner all of which are crawler-mounted. The cutting heads are made up of drums with picks together with a built-in loading and conveying system.

NCB/Dosco In-seam Miner. This a hydraulic in-seam miner as shown in Fig. 16.

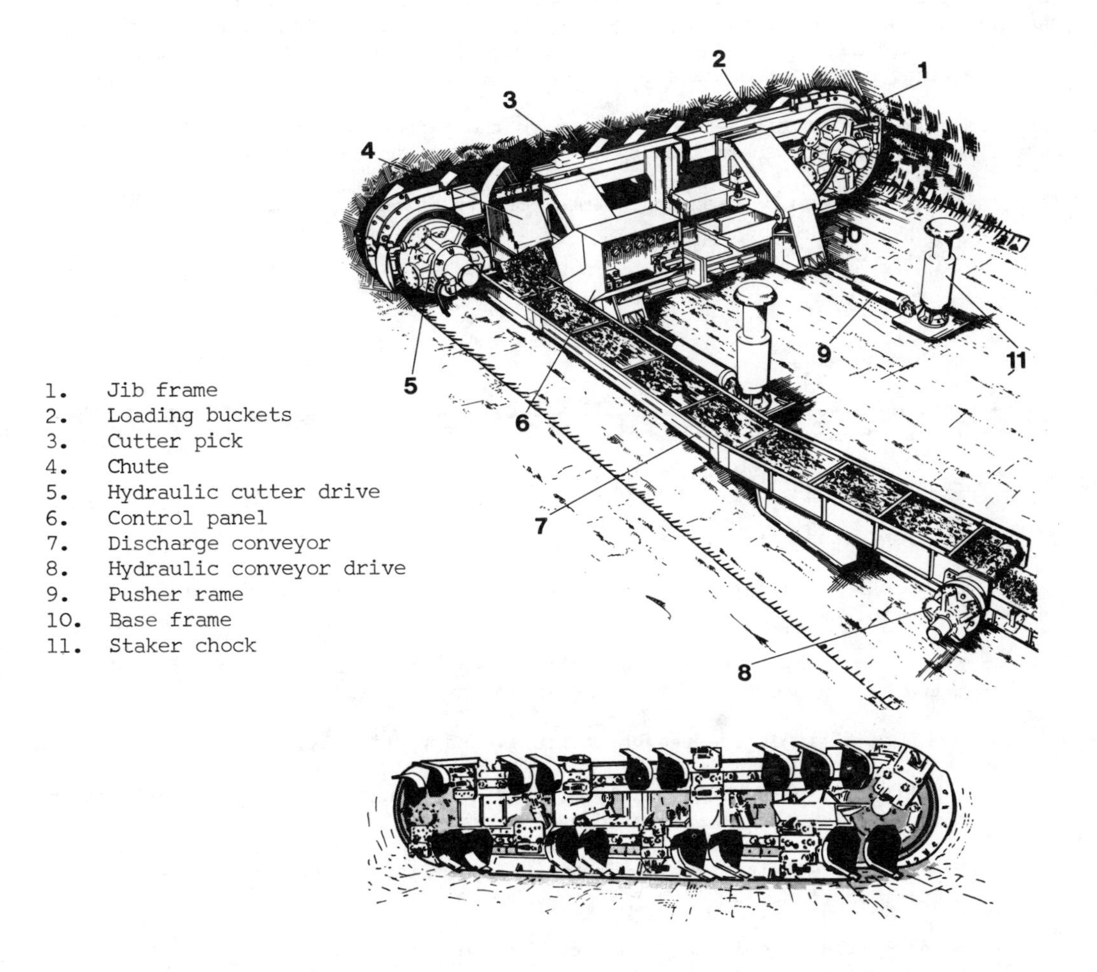

Fig. 16 NCB/DOSCO In-seam Miner developed at M.R.D.E.

The machine is based on a simple arrangement of cutting picks mounted on carrier plates attached to an endless strap-link chain running on guides mounted on a jib frame and driven by a sprocket at each end of the jib. Each sprocket is driven by a hydraulic motor. Loading buckets attached to the same chain deliver the cut mineral, through a chute mounted on the jib frame, to a hydraulically powered discharge conveyor.

The jib is attached to a base frame by means of two pivots - sometimes three on wider machines. Pushing rams, anchored to powered clocks advance the machine into the face.

Vertical steering is effected by raising or lowering the jib relative to the base by means of hydraulic rams mounted in the pivot brackets. Horizontal steering is by pushing ram, attached to a pivoted pad located at either end of the machine. Cutter chain tension is maintained by a hydraulic tensioning ram. Hydraulic power is supplied by a 90 KW power pack. Spray blocks are mounted on the jib for dust suppression.

The height of extraction for machines taking 1.07 to 1.27 metres is varied by changing the position of the pickboxes on the pickplates. The same method is applied to machines designed to extract between 1.27 and 1.45 metres. Conversion from one range to the other is brought about by modification of the jib assembly.

There is also an electric version similar to the hydraulic miner except that the cutter chain sprockets are driven by electric motors up to a maximum of 48 KW.

This in-seam miner can be used in a number of situations:

- In-seam roadway drivage.

- Face drivage.

- Stable cutting.

- Coal heading in thin seams followed by mechanised ripping and packing of the stone made during the roadway formation.

It is versatile in that it:

- is simple and safe to operate.

- cuts coal and relatively soft thin stone bands up to 34 MPa compressive strength.

- has a low pick/chain speed - 7 metres per minute with 18 picks on a 4.27 metre centre machine - resulting in low dust levels.

- can be positively steered horizontally and vertically.

- readily dismantles into easily movable components and can be quickly assembled - say in the time covered by five shifts.

- can operate within a small prop free front distance under a variety of support systems.

- needs little maintenance and cuts a strong smooth profile giving good strata control.

A number of low-height roadways have been driven in various British coalfields in seam sections between 1.22 and 1.45 metres.

Large roadways have been driven by extracting the coal with an in-seam miner and cutting the stone rip using a ripping machine or by firing and loading. A 9.75 metre centre machine was used in this way at one colliery to extract the coal. The ripping dirt was cut mechanically and packed into the packhole made by the machine.

Dintheaders. The Dosco dintheader first came into use some ten years ago. Crawler-mounted and electro-hydraulically operated, it cuts a square profile up to 7 ft. 2 ins. high using picks mounted in cutter chains. The cutting head is 5ft. wide and the cut debris is carried by the pick mat into a built-in chain conveyor. The whole machine weighs 13.5 tons being powered by a 120 h.p. water cooled electric motor. Their use is restricted mainly to British mines.

Roadheaders. This machine is produced by a number of manufacturers in the mining world and is illustrated in Fig. 17.

Fig. 17 Road Heading Machine

The Russians were first in the field here, with their PK3 introduced during the early 1960's. Roadheaders are versatile being able to cut coal or stone of medium strength. They are also able to cut an arched or square section roadway.

This type of machine is easy to transport and is well suited to the shorter lengths of drivage when cross gates and junctions are needed.

Essentially, these machines consist of a gun barrel type of boom mounted on a self-propelling framework probably with crawler tracks. The boom may be telescopic operated hydraulically enabling the cutting head at the end of the boom to reach the edge of the roadway being driven.

The debris produced is loaded simultaneously from the front of the machine, using gathering arms, onto a central conveyor or an arrangement connected indirectly.

The cutting heads are attached to the boom either on the same axis as the boom or on a shaft at right angles to it.

A number of countries produce this type of machine.

Austria. There are two machines made by Voest. Both have gathering arm loaders delivering on to a centre chain conveyor and are crawler mounted. A later version has a cutting head rating of 225 KW.

Britain. Anderson Strathclyde offer a range of machines. These include:

- Gathering Arm Loaders which are tractor mounted with hydraulic controls for forward and reverse movement conveyor and gathering head clutches, gathering head and discharge conveyor jib slewing and conveyor chain tensioning.

- Roadheaders with two electric motors for cutting head and hydraulic power pack, gathering apron and conveyor system and electric control gear.

- Boom miners for driving roadways down to 1.67 metres but equally up to 3.42 metres. Being short it is useful for road junctions. It is crawler mounted exerting low ground pressure.

Hungary. This machine is crawler mounted with a pair of gathering arms on the front apron which load the debris on to a central chain conveyor for disposal at the back of the machine.

West Germany. Demag produce a range of machines, mounted on walking bases with a double set of props wedged between the roof and the floor beams. This design provides temporary roof support over the machine and permits the setting of permanent supports, outbye the machines, as cutting proceeds. This speeds up the effective rate of drivage.

Both machines have axially mounted cutting heads. The debris is removed from the front of the machine by a flight loading chain surrounding it just above floor level.

The VSI is designed to work in headings measuring between 1.2 to 2.8 metres high and from 3.5 to 5 metres wide, cutting coal or soft stone.

This machine is shown in Fig. 18. It can be used in thin seams under adverse geological conditions and in inclined and dipping roadways. The main features of the machine are the seven hydraulic cylinders which form the clamping system and a canopy which clamps it between roof and floor, so that during cutting the machine does not rely on weight for its stability.

The machines have been used successfully on gradients of 45°C and steeper. They are also suitable for operation in side-dipping roadways, the tendency to side-

Fig. 18 VS1 Header

slip being countered by steering the machine to the rise.

It can also be dismantled into small packages for easy transport. This speeds up movement from one site to another making it economical for short drivages.

A later version, the VSS is suitable for use in rectangular and arched tunnels with cross sections of 12 to 30 square metres and capable of cutting rock with compressive strength of up to 800 Kp per square centimetre.

Paurat builds machines very similar to the Dosco type. The later type is fitted with a cutting head of 200 KW.

Eickhoff produces crawler mounted roadheaders with gathering arm loaders delivering to a central conveyor. Cutting heads are mounted radially or axially with ratings up to 200 KW.

Westfalia - Lunen offer a range of selective cutting machines from powerful mining, heading and tunnelling machines to efficient boom cutter loaders. All are crawler mounted. The main features are the cross-axis ripping-type cutting heads, with ratings up to 300 KW.

Machines in the WAV 200 series are provided with loading and conveying assemblies which discharge onto a bridge belt conveyor. Dust control equipment using water spraying and extraction systems are fitted.

<u>U.S.S.R.</u> produces machines with axially mounted cutting heads at the end of their

respective beams. Propelled by crawler tracks the debris is loaded by a cantilever flighted chain conveyor which encircles the machine. Gathering arms of the latest version feed onto a centrally placed chain conveyor. The cutting head rating is 90KW.

Full-Face Tunnelling Machines

These have not been produced successfully in Britain nor are they used to any extent. One is installed at Durham but it is not capable of placing the supports immediately behind the head.

A full-face 6 metre high speed tunnelling machine should achieve 2 to 3 metres per hour including erection of the supports when used to construct access roads and main roads.

In Germany they form a link and concentrate output providing long main roadways between different mines. Shafts are introduced at convenient points to get men to the coal face more quickly. Thyssen and Siemag of Germany both market this type of machine. However, it is thought unlikely that 6 metre heads will be used in Britain. The method depends upon the economies and often it is cheaper to use other methods including men, as well as being safer. Long runs are an essential feature for economic full-face tunnelling.

Hydraulic tunnelling. As has already been described in Chapter 3, rock can be cut by means of high-pressure water jets. The pressures needed are exceptionally high requiring a corresponding installed horsepower which is not practical for coal mining. Problems of safety also result. A reduced power requirement might be possible through the use of a pulsating jet, a method discussed in the " Mine of the Future ", based upon experiments with this end in view.

Campacker. This is a face end system developed for the mechanical cutting of stone and packing the debris.into the sides of the roadway. The campacker was designed to form part of such a system. It consists of a number of quadrants each vertically pivoted and mounted within a protective box. The layout is featured in Fig. 19.

Each cam and box form a self-contained unit weighing approximately 400 Kilograms, suspended from the rear of the face supports in the packhole area. The cams make too and fro through a 90° arc, their movement being controlled by means of a hydraulic ram mounted within the cam box and capable of importing to the cam a tip load of about 40 KN. Power is supplied to the ram from a power pack situated in the gate, or from an auxiliary supply on the ripping machine.

Debris is delivered by a feed device on to the floor of the packhole immediately against the goaf side. As the height of the pile increases the first cam can be operated to sweep the material along the packhole to the next cam, which is mounted on the adjacent support. From here, the material is passed to the next cam and so on. When all the cams are in operation the material is conveyed along the packhole and progressively lifted and pushed against the roof and the face of the previous pack. The process continues until the operation is completed.

The campacker can be used in conjunction with various systems of ripping provided that a feed system is installed to transfer material from the roadway into the path of the first cam. It can be used on either side of the road, or both, with a ripping machine and an NCB/Dosco in-seam miner. The latter provides sufficient coal from the side of the roadway to allow all the ripping debris to be packed.

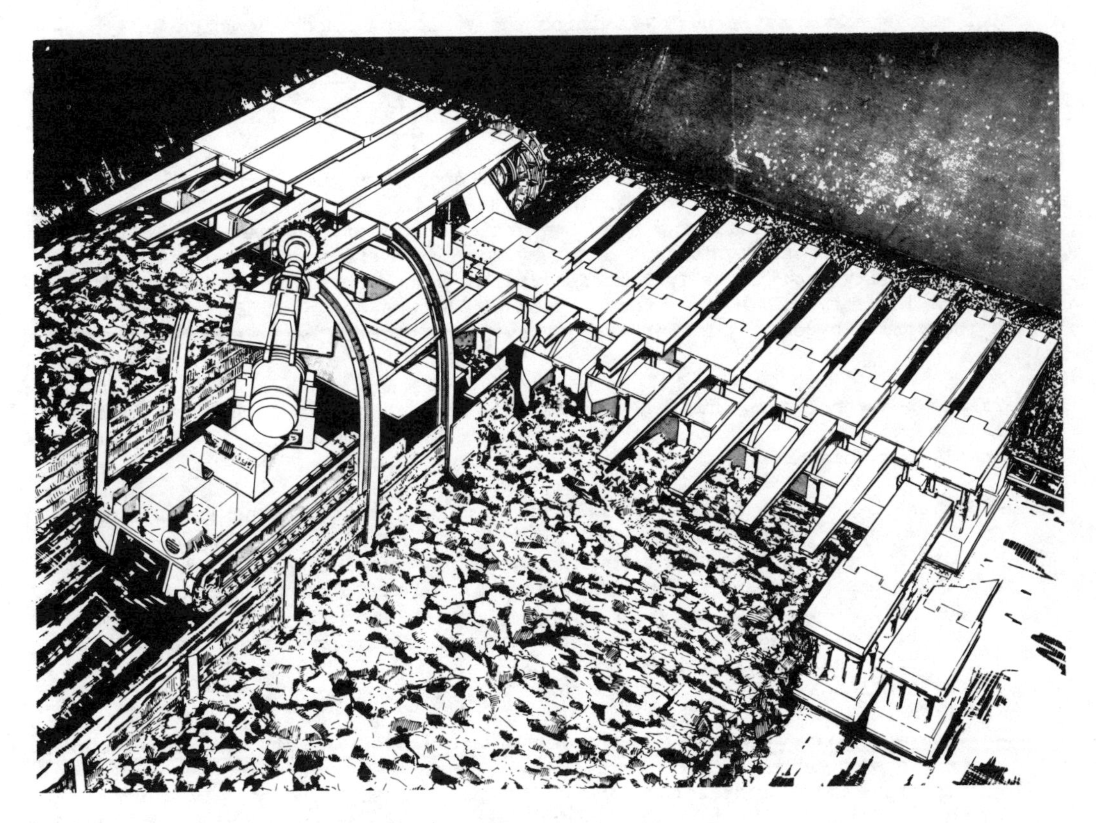

Fig. 19 The NCB Campacker developed at Bretby

The machine can be used in seam sections from 1.25 metres to 2.5 metres.

Mechanisation in Coal Recovery

Two methods are generally used to work underground coal deposits: bord and pillar and longwall mining. They differ mainly in the amount of work for road development and pillar extraction and the possibilities of mechanisation and concentration of production.

Extraction of coal from a pillar between two roads is practised in both cases. With bord and pillar mining and subsequent extraction of pillars - dependent on the size of room and pillar widths - some 30% of part of a deposit is exposed through development and roads. With longwall mining only 5% is involved in this way. A diagramic layout of bord and pillar is to be seen in Fig. 10 while that of the longwall method is featured in Fig. 11.

During the past decades both methods have been mechanized to an increasing extent and have reached a high technical standard. Many attempts have been made to introduce the bord and pillar system in Europe and the longwall system in those coal fields where the bord and pillar method was established. However, only the longwall system succeeded to any degree, where the bord and pillar system was traditional. At present there are around 80 longwall faces in the American, 5 in the

Australian, 20 in the Indian and some 20 in the South African coal fields.

The coal mining industry is busy examining the various methods by which unit output and productivity can be increased. For most of the countries, where coal production will be stepped up in the near future the longwall system is not, as yet, extensively used. It sintroduction could provide some of the increased tonnage at present being sought.

In the USA today less than 4% of the total underground output is mined by longwall systems but there is growing interest in this method. It is now quite apparent that the increase in continuous mining production has been achieved by the introduction of more and more machines and not by the improvement of productivity. In fact, productivity from these systems has fallen and productivity underground in the US coal industry has dropped from a peak of 15 tons per main 1969 to 8 tons, before the strike began. Similar trends have already been experienced in British mines.

Comparisons between the two mining methods show that the fully mechanised longwall system is capable of producing a further considerable reduction in the cost of mined coal, for these reasons:

- It enables a high degree of concentration of working,a beneficial impact on transport, ventilation, man-riding, power supply, etc., Annual output of 2 to 3 million tons requires a yearly extraction of coal over an area approaching one square kilometre. Whereas continuous mining bord and pillar equipment only occasionally achieves a unit output of 1000tons/shift, the longwall system is capable of exceeding 2000 tons/shift. On average, the longwall system has a daily output of about 5 times that of the continuous miner.

- It assists roof control and enables higher support safety with lower support costs, under difficult roof conditions. This is of considerable importance with increasing working depths. With increasing roof pressure the relative widths of rooms and pillars become changed to the disadvantage of pillar winning so that an increased recovery of coal cannot be achieved.

- It results in a higher proportion of actual working time because with the face equipment working in one place, continuous mining results. It does not have to be mobile and therefore can be built very robustly so that maintenance and repair costs are lower.

A recent survey of the two systems by an American Consulting Engineering Group reported that a build up of output at a mine to 2.0 million tons of saleable coal requires: either the installation of 23 continuous coal faces or 3 longwalls and eight to ten continuous faces.

Ventilation needs at the mine would be 200,000 cubic metres per minute for the bord and pillar mine or 110,000 cubic metres per minute for the longwall mine.

The high initial cost of longwall installations under US - conditions of some US-$ 5.4 mill. as compared with US-$ 1.3 mill. for continuous miners is outweighed by the substantial savings that are attributed to longwalling estimated at approximately US-$ 2.4 mill. per year.

The latest trends in longwall mining mechanisation exhibited at the 1976 International Mining Exhibition in Düsseldorf and at the 1977 International Mining Exhibition in Birmingham showed that considerable attention is being paid to increased unit production especially increased safety.

Longwall potential. The benefits of this method have been clearly indicated, but even so there are differences between the British and U.S. approach.

Retreat longwall mining is applied almost totally in the States. It is likely to continue as the main method particularly where medium to thick seams exist.

Advancing longwall will probably make progress particularly where seams are thin and also where high quality metallurgical coals, under deep cover, need to be worked. The reason is tied to the performance of the Continuous Miner, but it is the declining performance of these machines which is causing the move to longwall in general.

One of the essentials to Retreat Longwalling is the necessity for development of necessary replacement faces in sequence with the extraction rates being achieved in the face itself. Further, the requirement by Law to have multiple entries driven for each unit could make mine operators take another look at Advancing Systems with their low development drivage requirements. However, the introduction of single entry which might be possible under the law could give a further impetus. Development with monolithic packing systems may also help in this direction by reducing some of the delays experienced at the ends of Advancing Longwall faces.

Generally speaking, longwall units in the U.S. tend to be shorter than those found either in Britain or Germany. Face lengths of 450 to 500 ft. are common, although there is a tendency towards longer units, probably up to 800 ft.

The capital cost of longwalling in the U.S. is much higher than in the U.K., although British longwalls are generally longer than those found in the States. From figures at 1975 prices the average cost of longwalls in British Midland coalfields was about $950,000 while the cost per foot run was $1600. At 1977 prices the cost of equipping a 500 ft. longwall in the U.S. was $4.5 million or $9000 per foot run, while comparative British figures had risen to about $1.25 million or $2093 per foot run.

The U.S. mining industry is heavily dependent upon British and German research and development as well as European manufacturers of equipment. It is in this respect that the National Coal Boards Mining Research and Development Establishment (MRDE) is vitually unique, there being no real counterpart either in Germany or the States. The comprehensive nature of this co-ordinated programme can be seen in Appendix B where the main projects are listed. The German programme is also to be found in Appendix H, but this is carried out mainly by industry funded to varying degrees by the State as shown in Chapter 4.

Coal-Getting Equipment

There are three main constituents to the coal getting process - Face Supports - Shearers and Ploughs - Belt Conveyors.

Longwall Face Supports

The key to modern longwall technique with fully mechanised face operation was the introduction of powered self-advancing support during the sixties. Nowadays, the majority of longwall faces all over the world are equipped with powered supports - a satisfactory roof support system is essential in mechanized coal winning both for the safety of the miners and the efficient functioning of the coal-getting machines. These supports consist of two to six hydraulic props mounted in a floor base and topped by roof bars of canopies. The roof bars are cantilevered ahead of the front row of the hydraulic props - a distance sufficient to extend across the width of the face conveyor. There is a tendency towards one-web-back system (or Immediate Forward Support) in the shearer faces, whereby before the cutting operation the supports are kept a web distant from the conveyor. Four basic types of self-advancing supports are available - Frame-Support - Chock-Support - Chock-Shield-Support - Shield Support.

They differ mainly in their basic stability characteristics, roof coverage and method of advance. While the frame and chock types must be under load to gain stability for roof support and are sensitive to horizontal forces, the chock-shield and shield types have a built-in stability resistant to horizontal forces.

Two methods of advancing the support units are employed. The first incorporates the advancing cylinder in the support unit itself making it self-propelled (double or triple-frame support), while a separate "pushing ram" that can be powered according to conditions, is used to move the face conveyor after coal getting. With the second method the advancing cylinder is linked between the face conveyor and the support unit (chock and shield support) also serving to move the conveyor or the supports.

Capacities of individual props vary according to local conditions, designs with capacities up to 220 tons being in use. By means of built-in safety valves set at pressure of 300 to 500 bar, the props yield slightly to equalize loading along the face or compensate for increasing sag in the roof. The load setting of the props is obtained by pump units operating at pressures of 250 to 315 bar, ensuring that the props are pretensioned at 60 to 80% of their yield load.

The development of the lemniscate linkage at the rear of the shield is the most significant development in support technology over the recent years. The latest design trends in powered supports are the shield and chock-shield supports with pressure-initiated batch control using the components of the single-lever adjacent control system and additional valves for automatic sequence. In this way, any number of units can be advanced automatically by means of remotely controlled hydraulic impulses from a control panel on or off the coalface. One other promising development for powered supports in plough faces is a bi-directional remote control system with an outrigger steering system for horizon adjustment of the Gleithobel plough.

To hold a face at a constant and high level of production demands that the face equipment must be fully integrated and adjusted. Examples of peak performance trials carried out in modern faces indicate that production figures of 7,000 to 10,000 tons/day, district OMS of more than 40 tons/man-shift and area gains of face advance of more than 3 square metres per minute of machine running time are possible.

The German industry has been particularly successful in selling overseas, especially in the U.S. where some 20 sets of 2 leg Caliper Type Shield Supports have been sold over the past three years, while Klockner Ferromatic Shield Supports were sold to South Africa around 1975. Westfalia Lunen are also particularly active.

Japan has developed a 4 leg Chock Shield sold to Australia recently.

Russia offers less information, but reports indicate that Chock Shields and 4 leg Shields are in use, while Shield Supports seem to be gaining acceptance world wide.

Britain is a leading supplier, Dowty & Gullick Dobson being particularly well known. However, development in Britain is of more recent origin, impetus being given in 1972 when a Belgian group sought co-operation with Dowty in the development of a new heavy duty support. In 1974 South Africa showed interest in the Dowty 450 ton Chock Shields which gave a good account of themselves.

In 1975 Dowty were asked to design a 2 leg Caliper Shield to penetrate the American market, incorporating the lemniscate linkage now employed universally by Germany. In 1976 NCB ordered a full face of 4/300 ton Chock Shields which have demonstrated good roof control without flushing problems and the ability to negotiate roof cavities without having to stop and timber on top of the supports.

Other requests included one for a Shield Support which would close down to 0.65 metres - an order for 50 resulting.

Fig. 20 illustrates a shield support suitable for medium seams.

Increased safety in mining operations is provided by the Gullick Dobson advanced technology mining (ATM) powered roof support system. They have recently announced a remote control unit which permits the positioning of a number of hydraulic roof supports along a longwall face to be controlled remotely. It has the approval of both NCB and the Inspectorate.

Shearers and Ploughs. Mechanical coal getting in longwall faces is achieved in two ways.

1) ploughing by means of high-speed ploughs for flat seams and ram ploughs for steep seams.

2) cutting by means of a single or double ended drum shearers.

Recent developments of the former have concentrated on the design of ploughs with lower friction losses, plough blades similar to shearer picks, increased haulage during bi-directional cutting, mechanical or hydraulic adjustment of the plough cutting height, automated reverse of the plough at the face ends, digital display of plough position in the face including face alignment, elimination of stable holes, multi-speed ploughing and ploughing with overtaking speed. Here contrary to usual ploughing methods, the speed of the plough is higher than that of the conveyor.

Developments in shearer loader design have continued the trend towards higher-powered machines with one or two ranging arms. These shearers have taken the place of the orthodox types with rigid drums. The maximum installed power of drum shearer loaders is now 600 kW or 800 HP.

Most available machines are of the positive web type with seams varying from 0.75 metres to 5 metres. Cutting speeds vary considerably according to local conditions but they generally range between 2.4 and 6 metres per minute.

Machine power is generally within the range 65 kW - 80 kW for thinner seams while 150 kW is most common in thicker seams. There is a growing need for 200 kW in thinner seams and 300 kW in thick seams, although it should be said that there are a number of recently developed Double ended Ranging Drum Shearers at work with the facility for fitting two 400 kW motors on one machine.

Most Shearers to be found in Britain are probably single motor, single cutting drum, with spiral vane, being mounted on a fixed or ranging gear head. The under frame of the ranging drum versions would be non steering or roll steering on the fixed head type of machine. Haulage is provided by 18 mm or 22 mm round link chains, with pulls of from 10 to 20 tonnes available. The machine is conveyor

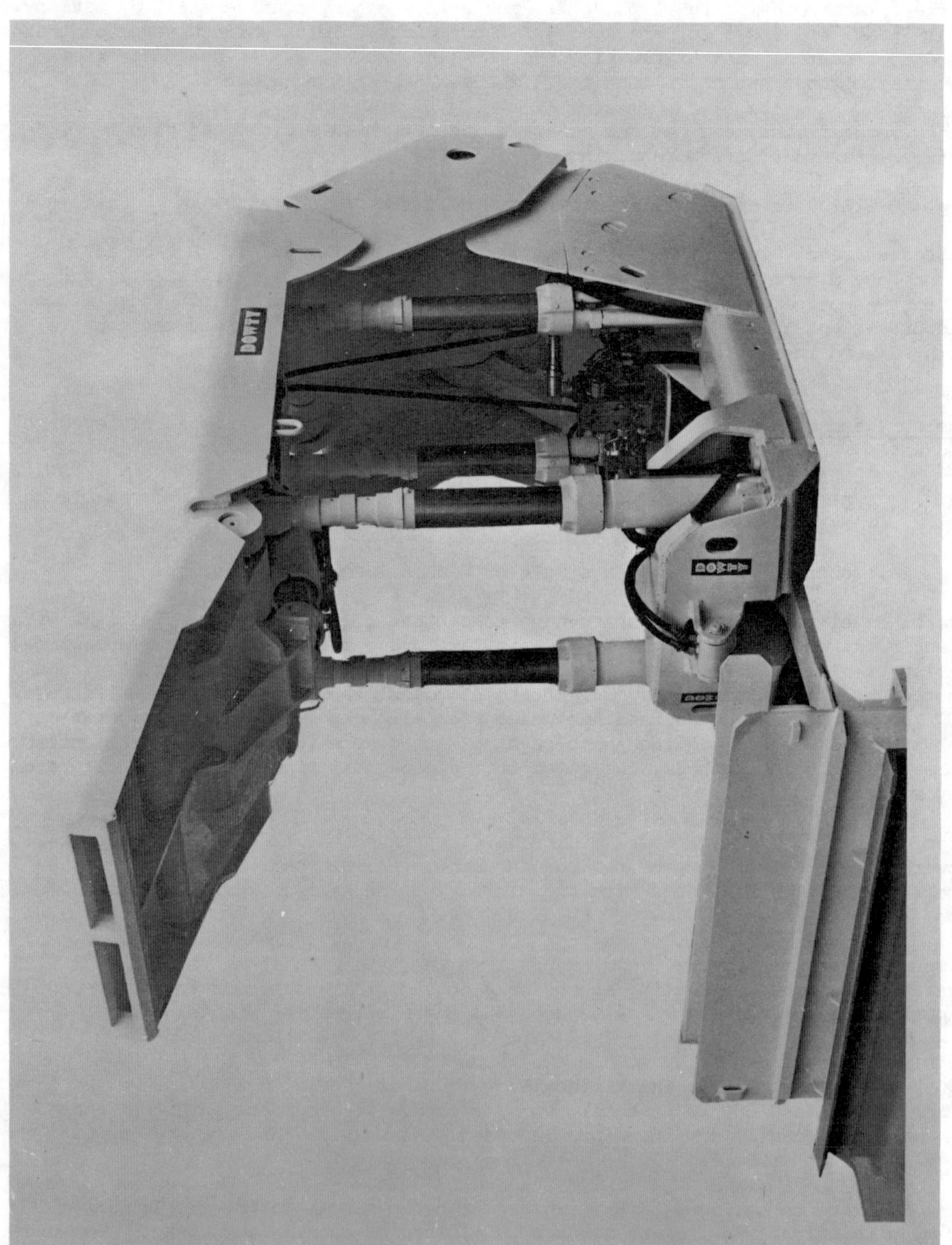

Fig. 20 Shield support for medium seams

mounted.

Double ended ranging drum Shearers are usually fitted with a single 150 kW or 200 kW electric motor with a haulage pull available of 20 - 27 tonnes using 2 mm round linked chain. A spiral vane type cutting drum is used at each end of the machine. Roll-steering is usually fitted, while an increasing number of machines are fitted with mechanical haulage gear cases instead of hydraulic. This change makes possible a choice of fixed cutting and flitting speeds.

The British Jeffrey Diamond Maximatic Shearer is shown in Fig. 21. It contains a number of interesting features now accepted as standard.

Fig. 21 BJD Maximatic Shearer with powered supports

Unit construction with dry joints.

Non handed arrangement of haulage and motor.

Simple re-handing of cutting units.

Goaf-side accessibility for easy maintenance.

External circuit condition test points on the haulage unit.

With correct choice of under frame and cutting unit this shearer can operate in a range of seam sections from 34 ins. to 12 ft. thick. It can be fitted with hydraulic or mechanical haulage and also adapted to be operated by remote radio control and become involved in monitoring operations.

The vital need for the elimination of stable holes at each end of the face has resulted in the development of long-reach cutting booms positioned outboard of the Armoured Conveyor to stretch beyond the conveyor drive frame.

The NCB's Mining Research & Development Establishment developed such a boom in conjunction with Anderson Mayo Ltd.

The first was installed in a 1.1 metre seam, to eliminate the tailgate stable, over a short eight-tooth sprocket AFC return end frame and also to cut down a further 1.1 metres of stone over the width of the roadway. The next seam to be tackled was 0.9 metres, while the third eliminated a maingate stable in a 1.25 metre seam.

Floor-based shearers. This type of shearer is manufactured by Anderson Strathclyde, British Jeffrey Diamond and Eickhoff. It incorporates the best features of the conveyor mounted shearer-loader in a double-ended machine which operates within and behind the buttock dug out by the cutting drum.

It has a powerful motor - up to 270 HP - standard hydraulic or mechanical units giving high pull via a 22 mm chain, ranging booms at each end to provide pitch steering and horizon control, as well as roll steering controlled by built in jacks and stabilisers.

The Anderson machine being off the conveyor offers full advantage of the reach which the ranging booms possess, the machine being easily diverted past the AFC drive heads.

All three machines have proved their ability to eliminate stables.

The floor-based shearer offers a number of advantages.

- a 30 in. web can be achieved within the 2 metre prop free front. Other webs are possible.
- the double-ended machine provides full seam extraction working bi-directionally; the facility for stable elimination at one or both ends of the face with single machine operation.
- good horizon control using pitch and roll steering.
- loading characteristics are good and clean up effective.
- improved coal clearance through the control bridge compared with conveyor mounted machines particularly in thinner seams.
- high rates of cutting and loading with minimum dust.
- hollow shaft ventilation with pick face flushing systems available on Anderson Strathclyde and BJD booms.
- makes the inaccessible accessible.

Rack-a-Track

The incidence of power loader stoppages on mechanised faces due to shearer haulage chain breakages has tended to increase over the years as higher horsepower machines and shearers for stable hole elimination, have been employed. This results in both lost production and a risk of loss of life.

A common rack for all types of machine has therefore been devised - Anderson Mayo, Eickhoff, BJD and SAGEM so that interchangeability is achieved. Currently all seam sections from 3 ft. upwards can be served by this rack.

The principle involves the engagement of a chain in a series of reciprocating pins, the peg lifting with the machine as it takes up the vertical tolerance in the trapping ensuring that the register between peg and chain is the same under all circumstances.

Once again the NCB MRDE has taken an Anderson Mayo shearer loader equipped with side haulage and fitted track reactive haulages. In this instance the haulage system consists basically of a rack and pinion with the drive unit bolted to the shearer. Sliding friction is largely eliminated by mounting the shearer loader on a roller underframe capable of reducing friction by one third of normal resistance of a sliding underframe.

The power loader is able to negotiate the maximum vertical and horizontal articulation of the conveyor. The main advantages include reduced operational costs, greater safety and the fact that more than one power loader can be operated on the same face conveyor.

Automatic Steering

NCB's MRDE has developed the Nucleonic Steering System already applied to fixed drum shearers and now being tested on Ranging Drum Shearers. The system requires that a thickness of coal be left above the machine which is continuously measured by a radio active source which in turn activates the machine steering jacks to maintain the agreed coal thickness and so the necessary cutting horizon.

Face Conveyors

Over the years there has been a considerable movement towards more robust Armoured Conveyors, capable of dealing with higher tonnages as well as carrying heavier machines than the 200 mm high conveyor. Work has been continuing using drive heads capable of 180 kW to meet the increased capacity. The best application for single centre strand, dual inboard centre strands and the more conventional double and triple strands chains is considered as well as the geometry of the drive heads to meet stable elimination needs.

German conveyors range from 632 mm (PFI: 500 mm chain centres to 871 mm (PFIII: 700 mm chain centres). The chains used vary in size and breaking strength from 41 tons (18 x 64 chain) to 113 tons (30 x 108) according to the Westfalia Lunen range.

The face conveyor is used for loading purposes. This demands high thrust. It can be moved by separate ram units or sets fitted in the base of the support units. They may be of fixed or variable pressure type, the latter enjoying a greater degree of flexibility. With ploughs the ram rarely exceeds a thrust of 2 tons, but with shearers this may rise to 8 tons.

The conveyor may also be used to carry the chock and shield supports and must therefore be capable of bearing stress of 30 to 40 tons which occur with heavy duty support units during an advance operation.

Transport Belt Conveying

It has been the custom for continuous belt conveyors to be the main means of transporting coal from the face but their application has been extended so that men and materials can be carried. There are however problems, since this is a means of assisting the passage of fire through the mine which must be considered carefully.

New belts were necessary using synthetic fibres such as nylon and terylene woven into fabrics, impregnated and covered with PVC, to provide a flame retardent belt of adequate strength. Plied or solid woven construction are used. The latter have the necessary degree of toughness coupled with flexibility in both directions necessary for mining needs. Modern techniques should enable belt conveyors of up to 1100 kW to be constructed.

For major surface and cross measure drift installations, steel cord belts, embedded in a matrix of rubber are being made flame resistant up to mining standards.

It has been necessary to introduce the module principle with regard to drive. Four transmission units applied on two driving drums are fitted at the head end. Alternatively a driving unit having three drums with a single transmission for each has been introduced since it solves problems of width including two transmission units on one driving drum. The exact arrangement appears to vary from one manufacturer to another but all follow the same general principles.

The important stage of transfer without loss between conveyor units is now being tackled. The movement of coal from face to gate usually includes a scraper chain loader with a fixed flexibility to the face and gate conveyors to ensure centralised transfer. Water, used to reduce dust to a minimum, tends to cause some of the coal fines to adhere to the return belt which can only be removed by the use of polyurethane, mild and hardened steel strips and flexible wire ropes, according to circumstances.

Belts used for carrying men must conform together with the general construction, to very high standards of safety - to be discussed in Chapter 11.

Dust Filtration

NCB's MRDE, has developed a dust filtration unit in which dust laden air is drawn into the fan where it is mixed with a finely atomised water spray. From the fan the mixture of dust, water and air is drawn through a removable filter panel supported by heavy back meshing and protected by a stainless steel screen. Dust is trapped and drained from the filter assembly to the action of water, most of which drains directly into a settling tank below the filter housing. Some of the water, in the form of coarse droplets, is discharged from the downstream face of the filter. This is trapped in a droplet removal section and drained into the settling tank. The water is recirculated.

Pneumatic Power Pack

Power packs for operating standard pneumatic tools underground have been developed using liquid nitrogen stored under pressure, not exceeding 150 psi, in a vacuum insulated container.

When a tool is connected, the liquid is drawn off boiled and superheated to within a few degrees of ambient temperature in a heat exchanger using heat from the atmosphere. The gas passes through an adjustable regulating valve to reduce its pressure through self-sealing connectors and on to the tool.

Monitoring and Control

No review of mining equipment and the way it can be used to recover coal would be complete without some reference to this important aspect. It has already been discussed by implication in Chapter 3 and will play a prominent part in the "Mine of the Future".

Taking the NCB situation there are said to be some 238 producing collieries, most of which have installations for controlling items of underground plant from the surface. Interest in general, Britain, France, Germany, U.S.A. and U.S.S.R. paying particular attention to this important field.

MINOS

This range of commercial mine monitoring and system controls was developed by NCB's MRDE, at Bretby. MINOS is a standard computer system with control console and operating system software designed specifically for colliery operations. The systems in a variety of forms can be supplied by a number of data transmission equipment manufacturers.

The objectives are to:

- provide a modular computer control system for collieries which can be installed now and yet be compatible with and capable of accepting future developments.
- design systems with a high degree of technical competence reliability and safety.
- obtain a common standard throughout the industry while maintaining utmost flexibility to accommodate local variations and needs.
- provide facilities for control adjustment of operating constants and selection of plant options - changes in local layout and plant characteristics - in a way which requires little training.

The range of systems include:

1) Production monitoring.
2) Management information systems.
3) Coal transport control and monitoring.
4) Bunker management.
5) Store and coal segregation and blending.
6) Switchgear monitoring and control.

7) Environmental monitoring.

8) Main pump supervisory control.

A number of these systems may be controlled from one point or computer system.

Equipment. A standard set of equipment forms the basis for each manufacturers system and includes:

- Standard control console with function keyboard and twin monitor screens. Colour is available.
- DEC computer PDP11/34 with floppy discs for rapid loading and data storage: 32 K basic memory expandable up to 128 K.
- Manufacturers own data transmission system with a range of defined out-stations.
- Print out for management and engineering reports.
- Remote display terminals for special purposes.

Production monitoring system. There is a standard package providing Data Transmission linking the gates to a surface controller providing automatic monitoring;

Face power loader	-	up to 3 machines
Face AFC	-	running/stopped
Face signals	-	lockout with indication of position
Stage loader	-	running/stopped
Face output	-	tonnage from belt weigher in gate

Power/current transducers coupled to trailing cables to monitor motor loadings.

Manual entry, Displays and Reports printed on demand are additional features.

Coal clearance control. The object with this equipment is to obtain automatic operation of coal transport systems comprising conveyors and bunkers. This will result from techniques for optimum network control including store/coal segregation, batching, blending and automatic implementation of bunker control policies.

Application will be very similar to production monitoring to cover status of each belt, conveyor delay summaries, data transmission system status and automatic alarms. It will also provide print outs of information vital to management.

Environmental monitoring system. Centralised automatic and continuous monitoring of all aspects of mine environment including highlighting abnormal and danagerous conditions forms the objective of this package to include:

Air percentage methane - BM1

General body air velocity - BA2/BA4

Methane 0 - 100% - BM2H

Drainage range suction

Duct velocity trips

Smoke detectors

Vibration

Pressure

Dust - temperature - humidity (under test)

Automatic handling of data from microprocessor - based tube bundle systems, are applicable for the following :

Methane - Carbon Dioxide - Oxygen - Oxygen deficiency - Graham ratio - Carbon Dioxide

Environmental status displays are also available.

The equipment reviewed in this chapter has been basically about mining. It hasn't covered every fine detail, nor could it possibly do so. Those interested in further application of remote control including systems employed and wireless can do no better than refer to the technical papers presented at the 1977 Birmingham International Conference on Remote Control and Monitoring Mining, which contain a wealth of information offered by those expert in this field.

Chapter 7

PREPARATION AND THE CONSUMER

Earlier in this study, the vital role played by efficient transportation was stressed so that the movement of coal from the face was such that recovery and preparation were co-ordinated.

In Polish mines, about 90% of coal transported is by conveyor. The rest is made up of locomotives, trackless vehicles and with small incursions into hydraulic systems.

Efficient transport systems will ensure higher production rates by reducing:

- delays to coal production
- number of people employed
- degree of loss of quality of the coal
- numbers needed for maintenance and in turn costs

Capacity will be increased correspondingly.

A number of traditional systems of transporting coal are in operation, belt conveyors first being introduced with longwall mining. Previously, tubes and ropes or chain were used. However, available power coupled with limited belt strength prevented higher capacity being attained. Fire resistent belts and improved systems have helped considerably, but even so there are still drawbacks and difficulties, delays resulting at a number of points - outbye - gate - stage loaders - bunkers - shafter drift - and other reasons. These delays account for a considerable proportion of total outbye hold ups. They have been found to be accounted for by:- mechanical failure - obstruction - communication faults - electrical failures - other influences.

It is considered that gradual change in the physical properties of the coal being handled, has contributed to some of the difficulties associated with belt conveyors. Power loaders result in smaller coal, while use of water to suppress dust, discussed more fully in the environmental section, have tended to cause the coal to stick to the belt, so preventing smooth transfer from one conveyor to another. It is the transfer problems which seem to have dominated the scene, many mine managements apparently taking the view that trouble free operation would not be achieved by remote control and therefore this advance in control should not be used, despite the benefits of certainty associated with electronic systems.

The crying need in the interests of reduced manpower is for a means of ensuring free movement from one conveyor to another so that the system can be truly automated.

However, good systems of communication still leave much to be desired, more comprehensive arrangements being needed. But the use of data link systems has improved the transfer of information as compared with the former direct wire method. The recent introduction of the mini-computer offers opportunities not available previously, to introduce remote control and the accompanying savings in manpower which improved electrical control and monitoring systems should now make possible. It is believed that once achieved, the following improvements should be attainable in a medium sized colliery.

- Reduction in numbers of men at conveyor transfer points - 75%
- Output per man shift improvement - 5%

Pneumatic Coal Handling

A pneumatic coal handling system basically blows coal from the pit bottom to the surface, through a pipe installed in the shaft. Pneumatic systems of this type will certainly be complimentary to the conventional winding method: they could well replace them. Another possibility would be to install the hydraulic pipe through a surface borehole and so eliminate the long haul to the bottom of the shaft.

Here, all underground operations are monitored from a control cabin sited near the bottom of the shaft, where fault finding indicators are used.

Run of the mine coal sized about 25.5 millimetres is fed from the underground conveyor into the rotary air lock feeder normally sited about 100 metres from the up cast shaft and then transported to the surface using a positive displacement blower employing around 700 H.P. at depths of 300 metres. The coal is then discharged to a collection cyclone after which it is taken by a conveyor belt to the washery.

Hydraulic Transport

An essential feature of hydraulic mining is that the seams should be pitched so that there is a gradient to assist natural water flow. This may vary from a few degrees up to very steep seams found in Germany and U.S.S.R. where in the case of the latter, 80° is not unknown.

With hydraulic systems, the coal is carried to the main hydraulic lift chamber, from the face, by water flowing in open flumes. A fixed volume of water is taken into the mine and later a fixed amount is removed. The quantity of the coal carried by the system, will vary dependant upon the number of coal faces operating and rate of production.

The main hydraulic lift chamber is located at the pit bottom, where slurry machines screen and crush oversized material down to the necessary grade. Coal pumps with or without boosters are then used to lift the slurry. Slurry pipe lines are con nected with the suction ends of the pumps on the surface to make the coal slurry

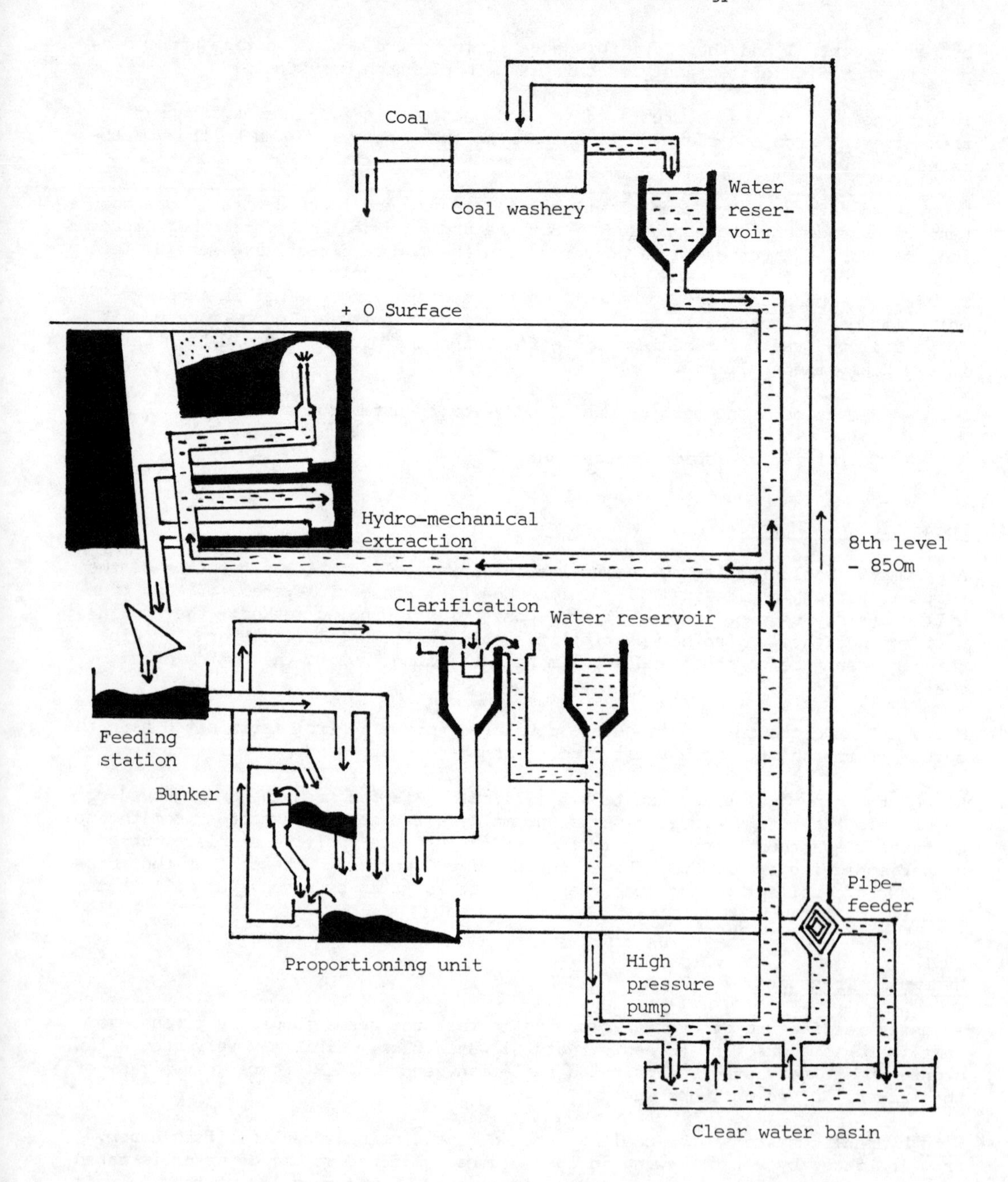

Fig. 22 Hydraulic mine layout

to the preparation plant. A diagramatic layout of this type of system is shown in Fig. 22 to include the arrangements for extraction of the coal by hydraulic means.

As was described in Chapter 3, over the past 20 years, the Soviet Union have developed a complete line of effective and reliable coal slurry pumps, high pressure pumps, feeders and ancilliary equipment for use in hydraulic mining systems. The Soviet Union probably has the greatest experience in this technique and therefore some statistical detail of equipment in use may be of interest.

Small one-stage coal slurry pump. This pump is for use in horizontal hydraulic transport systems. It is rated at 900 cubic metres per hour with a water head pressure of 90 millimetres.

Two-stage high pressure coal slurry pump. This pump has a throughput of around 1000 cubic metres per hour at 330 millimetres water head pressure. It has been used in hydraulic lift systems and main hydraulic transport systems.

Pumps alone do not provide the solution. Very considerable resources have been applied to the development of reliable pipe fitting for both water and slurry in hydraulic mining systems.

Couplings and fittings designed to ensure swift assembly and dismantling also form an essential and intergral part, as well as instrumentation to control the solid content. Valves have been designed for pressures at 900 psi.

United States. The experience of the Black Mesa Pipeline Co., is worth recounting. The pipeline is 273 miles - 439 kilometres - long and 18 inches - 457 millimetres - in diameter and carries the coal slurry from the Peabody Coal Company's No. 1 mine in North eastern Arizona to the Mohave Power Plant in Clark Country, Nevada. Its capacity is said to be 16,000 tons per day.

A major requirement of the slurry is that it should be homogeneous. This is achieved by batch blending employing four Denver agitators together with 49 ft. - 15 metres - diameter tanks with a depth of 45 ft. - 14 metres. Each is equipped with Denver axial flow 10 ft. - 3 metres - propellers. These are driven by 125 H.P. motors incorporating reducers.

The agitators are designed to circulate the slurry downwards through a central vortex and then outwards along the bottom and upwards along the inner wall of the tank. The specification is 53% solids by weight.

Pipe line flow is 660 tons per hour and forms the only means of transport. The system is therefore very dependent upon the reliability of the system.

Germany., They too have very considerable experience in hydraulic transportation. Siemag Transplan GmbH offer the three chamber pipe-feeder used at the Hansa mine. The system is installed near the shaft, enabling the movement of 5,000 tons of coarse grained coal a day, by hydraulic means, with sizes up to 60 millimetres from a depth of 860 metres and achieving a continuous flow to the surface.

The three chamber pipe-feeder works on the principle of a cyclic chamber lock system formed by long U type pipe curves. The diameters of the curves equal that of the delivery pipe line.

The cycle is begun when the chamber fills with a slurry operating under low pressure. Later water is introduced under pressure high enough to force the slurry through the delivery pipeline to the surface. Technical details include:

- Diameter of chamber 150 - 300 millimetres
- Length of chamber 200 - 400 millimetres
- Speed 4 - 5 metres per second
- Solid/liquid ratio 1:3 to 1:4

A central control system which is time operated controls the main valves and the pressure compensating valves ensuring continuity of flow despite periods of filling and emptying of the pipe chambers.

The general outlay of the system is to be seen in Fig. 23.

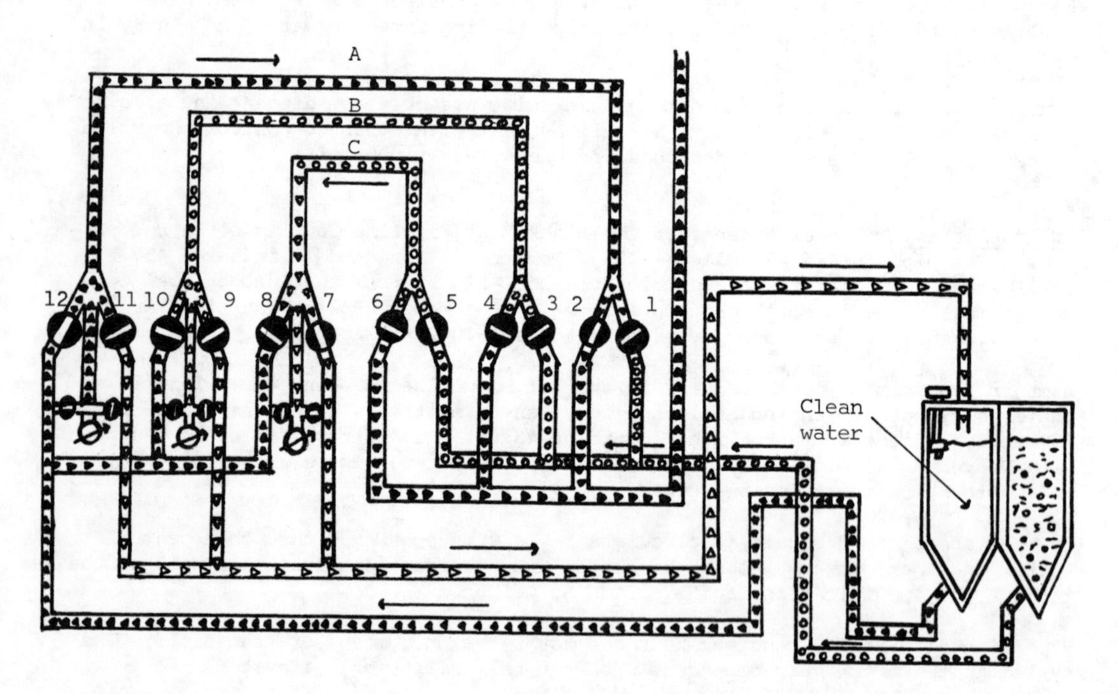

Fig. 23 the position in each part of the system is shown:

Pipe chamber A is emptying

Pipe chamber C is filling

Pipe chamber B is ready for conveying

The positioning of the valves help to trace this sequence.

Considerable advantages result from hydraulic transportation:

- Economic movement of coarse grained coals, 60 millimetres and above.
- Solids are kept away from the sensitive high pressure pumps.
- Hoisting by making use of available mine water as the transport medium, cuts costs.
- Shaft capacity is increased through installations of a delivery pipeline in the existing shaft; at reduced cost.
- Movement is continuous both in vertical and horizontal plans.

PREPARATION SYSTEMS

A large percentage of coal used today goes to electricity generation or coke manufacture; in Britain it is 75%.

When this type of pattern develops and becomes fixed it exerts a very considerable influence upon the way in which the coal is prepared and subsequently outloaded from preparation plants. This is in direct contrast to earlier times when some eight different grades were necessary. New plants tend to be designed for one grade only, adapted to the needs of the electricity generating industry. Other users must be supplied from smaller units which still produce minority lines.

This standardisation points to much simplier plant and automation with its reduction in manpower. One such plant in Britain now being completed, will need four men only instead of about twelve previously required. It has also meant that large capacity rapid loading bunkers can be installed with increasing use of liner trains to deliver one product to a bulk user. The bunkers hold up to 6,000 tons and are capable of loading moving trains at rates around 3,000 tons per hour. One train with one product results in a very simplified track layout system.

Characteristics of Coal

Modern methods of coal recovery not only produce problems of dust, but they also result in more small coals so that at most modern plants greater attention is given to the treatment of small and fine coals. But the treatment of small coals is not only very difficult, it is also extremely expensive. Variable feed rates up to 80 tons per hour of coals less than $\frac{1}{2}$ millimetre need to be treated, which means arriving at an acceptable level of ash too. As attempts are made and become successful to remove more coal from the roof of the seam so the amount of dirt has risen from 10% in the days gone by, to a current level of 40%. This may well rise to 60% in due course, particularly where "sandwiches of thin seams" become involved. Moisture contents will also influence plant design. And yet, today there is a shortage of "mining engineers" available to the firm supplying preparation plants - that is engineers with a mining background.

Preparation of Raw Coal

A consistent product is required. This means reasonably consistent raw coal without which there can be little hope of success.

Plant design therefore demands the incorporation of a system of blending and preparation of the raw coal before the cleaning process is carried out. Clearly automation is a vital ingredient.

Freedom from coal preparation problems depend upon feed: once this is right the problems go and costs become considerably reduced.

Two methods are used for achieving homogenisation

1) Layering in an open stockpile.

2) Systematic filling and discharge from conventional bunkers controlled through a computer programme to even out differences in coal size, consistency, moisture and ash.

The benefits of a blended feed to a washer are:

- plants can be designed for optimum feed rate and yield of the various products.

- maximum use of very expensive equipment reflected in lower costs per ton.

- components of blends, particularly washed smalls and filter coke are produced, in ratios which are complimentary, so avoiding surplus coke which must be disposed of separately.

- greater reliability of unit processes so avoiding hold ups and resulting in lower operating costs.

To date, Power Station Fuel only has been considered. But there are other markets requiring a more stringent seecification. These include:

- Coal as a source for Synthetic Natural Gas.

- Coal as a feedstock for the chemical industry.

- Coking coals requiring blending or treatment.

The process used will in the main be wet and involving:

- Filtration

- Froth flotation and multiple froth flotation

- Flocculation

- Sulphur removal by chemical or magnetic means

- Agglomeration techniques

- De-watering

- Homogenisation

United States

Approximately half of all the coal mined in the United States is cleaned in preparation plants. In the case of metallurgical coal the level is about 80%. As a result there are some 2000 plants with capacities from 100 to 3000 tons per hour. The trend is upwards, with train loading rates of 3000 to 4000 tons per hour delivering to 10,000 ton trains. Slurry pipe lines 1500 Kilometres long are on

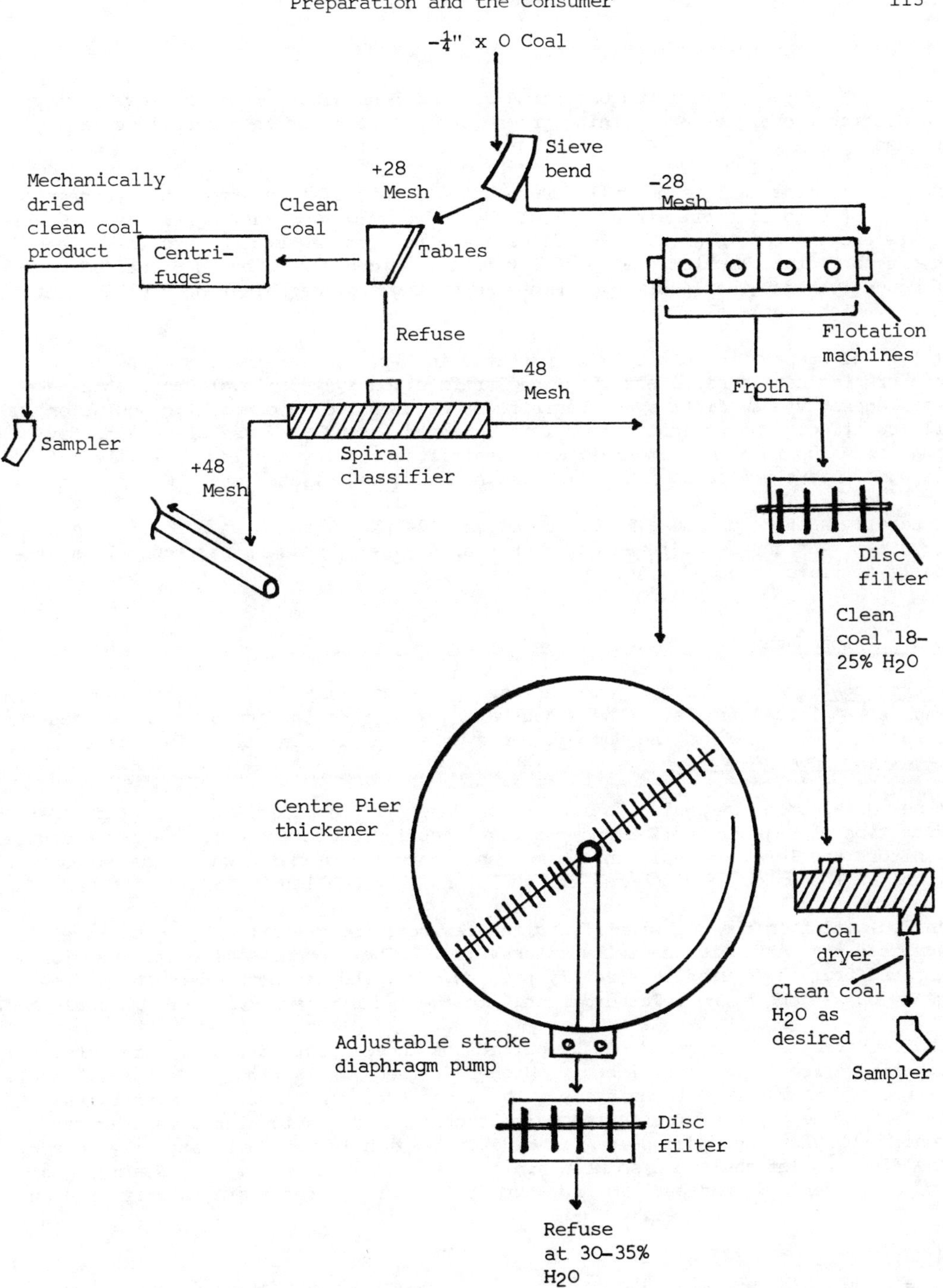

Fig. 24 Denver high efficiency processor

the way too for the movement of coal.

About 50% of the coal washed uses some form of Baum jig. However, low gravity separation techniques are gaining ground, when more accurate separations are needed.

Run of the mine coal is normally crushed to 150 millimetres after which screening occurs of 6 to 15 millimetres followed by hydrocyclones or shaking tables. Coarser coals are treated in jigs. The fines emerging from the cyclone are then cleaned by froth flotation which due to strict pollution control is being practised more frequently today than earlier. Fines resulting from mechnisation usually contain a very high ash level.

A typical preparation plant is illustrated in Fig.24. The problem of pollution has created considerable activity in clarification systems involving static rake thickeners, vacuum filters and centrifuges accompanied by chemical coagulation and flacculation. De-watering of coal is achieved through the use of screens for coal sizes up to 35 millimetres, with centrifuges employed below 15 millimetres. Vacuum filters are used with clean fins although centrifuges are now more common.

Generally surface moisture is of the order, 10-12%, although when lower levels are required 5-6% can be achieved using thermal dryers. This is particularly applicable to fluid type beds.

General Approach

Coarse coals. Jigs or heavy media vessels are normally used, the former of the Baum air-pulsated models. The Wedag Batac type, made in Germany, can be used for a range of coals, while the Humbolt Wedag gyratory air classifier introduces a revolving drum.

A new development in coal cleaning is the Rose type Baum jig incorporating radar detection of reject material and positive ejection of the reject. A radar device monitors the level of shale in the bed and controls special high frequency pulsations in the water to discharge the shale and the middlings.

The pulsations in the jig stratifying the raw coal are generated by compressed air, admitted through a high capacity rotor valve. They are dissimilar to and dissociated from those used to stratify the raw coal, the strength of the high freqency pulsations being related directly to the radar measurements of the shale bed.

Shale extraction is carried out by pulsations independant of and isolated from the stratification of the normal bed. Higher frequency pulsations in a special shale extraction chamber eject the shale without the need for air chambers or release valves. The depth of the shale bed collecting on the grid plates is measured continuously by the radar pulse "echo device" which is used automatically to vary the discharge of shale and so maintain the bed at constant level. Alarm signals indicate overload or departure from normal. Capacity for a single unit is up to 500 tons per hour raw coal.

Fine coals. Here again, air pulsated jigs, heavy media cyclones, Vorsyls, water cyclones and Deister tables are used. The British Mining R & D Establishment at Bretby is at present engaged on development of a Vorsyl Separator in which coal travels down the centre with the shale being centrifuged. This type of separation is extremely efficient. The Japanese "Swirl Cyclone" is another example of this increasingly necessary technique.

The Rose baum already mentioned can be applied to a wide range of coals, while other systems on similar lines if not so advanced are to be found in Germany and Japan - the Tabuc jig. The latter is capable of dealing with up to 1000 tons per hour and operating at densities of 1.45 specif gravity upwards, while above 1.50 sg efficiency is said to approach that of heavy media equipment.

Ultra fines. A number of froth flotation units are in use for dealing with sizes of less than 0.5 millimetres. This principle is fairly standard, except for units coming from Humbol at Wedag.

Other methods to attract interest include oil agglomeration, its main virtue being that it provides an answer to the quest for the recovery of more coal. It is capable of dealing with oxidised coal while the pellets produced have a low moisture content, making vacuum filtration or drying unnecessary.

Some mention should be made of vacuum filtration since it is used widely. Basically this method consists of the coal slurry being fed onto a rotating drum from which the air normally found within the drum is being exhausted. A continuous cake is formed much of the water being removed in the process. The method of discharging the coal from the filter drum after separation is a major design feature.

Water clarification is normally carried out with the use of thickeners. Here new types of equipment are being introduced. These include the Envirc Clear Thickener, the Lamella Thickener and the Deep Cone vessel incorporating a stirrer. The latter which is finding favour in Europe employs a steep angle cone, the stirrer releasing the water from the slurry.

The United States Bureau of Mines have developed a method of increasing the density of coal slurries using the passage of a DC current. This causes the particles to become negatively charged and as a result move to the anode sited at the bottom of the vessel.

Use of chemicals. In certains parts of the world where high sulphur contents cause considerable concern, attempts have been made to use chemical means for sulphur removal, although doubts must be expressed as to whether or not this will clean up the coal in the cheapest and most effective way. A number of processes have been recorded in the literature. These include:

- Hazen Magnex process using $Fe(CO)_5$ plus magnetic separation. 40% removal of ash is claimed and 100% in the case of pyritic sulphur.

- KVB uses dry oxidation followed by a caustic wash. Complete removal of pyritic sulphur and 40% organic sulphur reduction are claimed, but ash remains unchanged.

- Battelle Hydrothermal process uses a caustic leach which it is claimed will remove 100% pyritic sulphur and between 20 and 70% organic sulphur.

- Ledgemont oxygen leaching introduces oxygen, water and lime. Complete removal of pyritic sulphur is said to occur.

- PERC Air leaching system uses the same reagents as Ledgemont, with similar success, but in addition 20 to 40% organic sulphur removal occurs.

All appear to have attractions but will presumably be very expensive to operate.

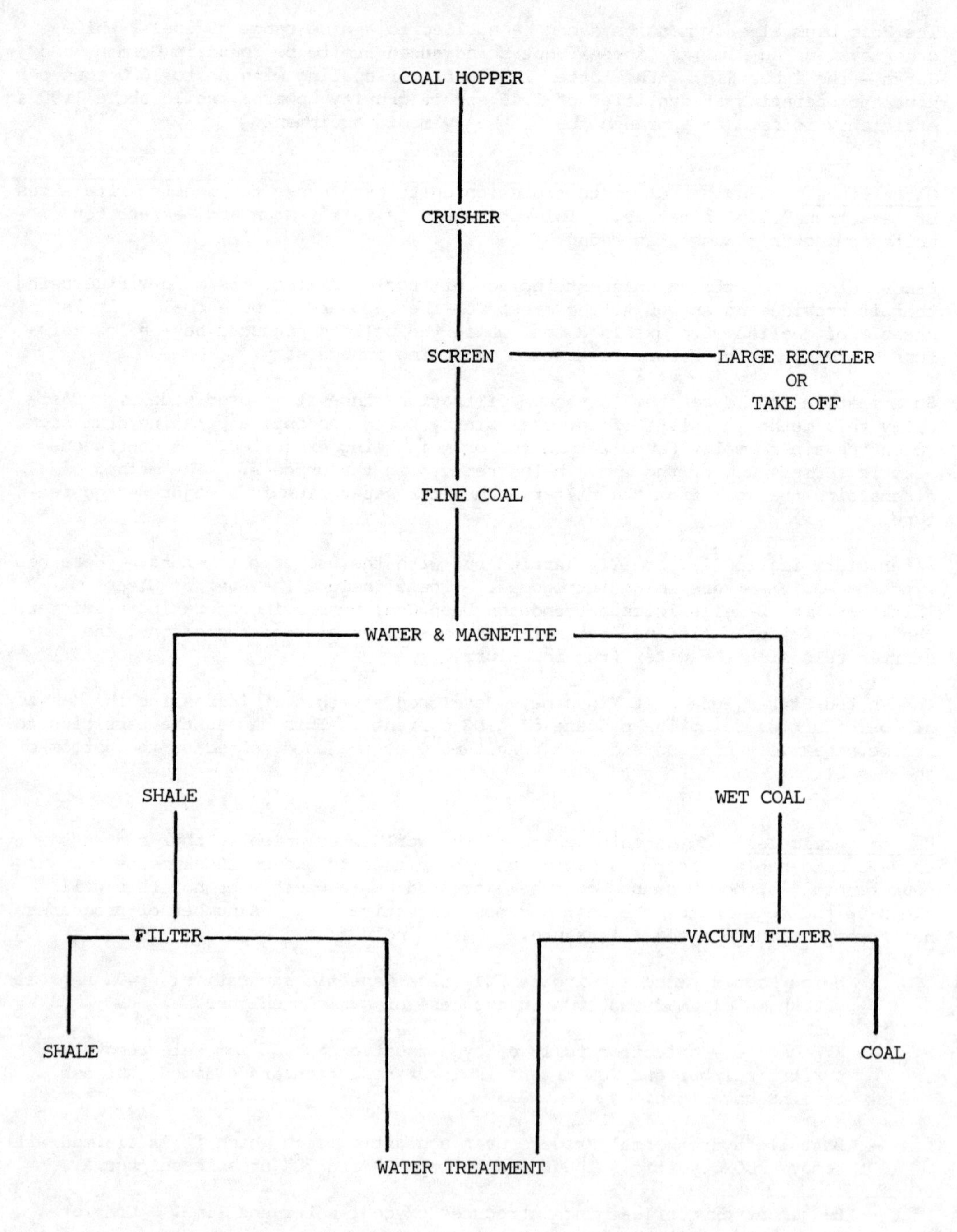

Fig. 25 Flow Stream for a modern coal preparation plant being installed in Britain

New Developments

The best of current preparation technology is to be found at the Prince of Wales mine Pontefract, in Britain, installed by Jenkins of Retford. The flow diagram is to be found in Fig. 25. Magnetite is added to the coal and water slurry changing the relative density so that the coal rises and the shale falls. The magnetite which is a mined ore is later recovered.

Looking at the components of a preparation system there are control devices available which make all the difference between smooth running and a homogeneous acceptable product, and problems. Sometimes the faults result from incorrect application of the controlling equipment or poor design of vital components such as the bunker.

A number of items of controlling equipment are now under development at the British Mining Research and Development Establishment which will be of interest to all engaged in mining operations. Some will now be discussed.

Nucleonic Density Gauge. This instrument measures the gravity of magnetite dense medium suspensions to an accuracy of ± 0.001 gms/cc. Market standards demand an ash and sulphur content which require rigid control. This gauge has therefore assumed considerable significance now that dense medium washers are being used more frequently to separate at low gravities.

Bretby Clarometer. This system has been developed to control the clarification of water used in closed-loop water circuits making use of high priced flocculants in a thickener to cause the solids and liquids to separate.

Variations in solids content and flow rate to the thickener feed require a compensating "variation dose" of flocculant for both effective and economic clarification of the water. The NCB Clarometer shown in Fig. 26 meets this need. It consists of:

- a sampler to take feed from the thickener after addition of the flocculant.
- a setting rate detector enabling the settlement time of flocculated solids to be measured.
- a control device to compare relative settlement times with standard or ideal time.
- a motorised valve operated by the control device to regulate dosage.

Bretby Ash Monitor. This device was developed jointly by NCB Mining Research and Development Establishment and the United Kingdom Atomic Energy Authority. It offers a rapid and accurate determination of ash content by continuous measurement of backscattered X-rays from a coal sample irradiated by a plutonium 238 isotope. Compensation for variation in iron content and ash are made involving the insertion of an aluminium filter of predetermined thickness between the isotope and the detector.

The monitor consists of a rotating table which forms a compact cake of coal. Above it is mounted the radioactive isotope assembly and radiation measuring system. The radioactive sources of the ash monitor bombard the surface of the coal cake penetrating up to 38 millimetres into the coal bed. Radiation is absorbed

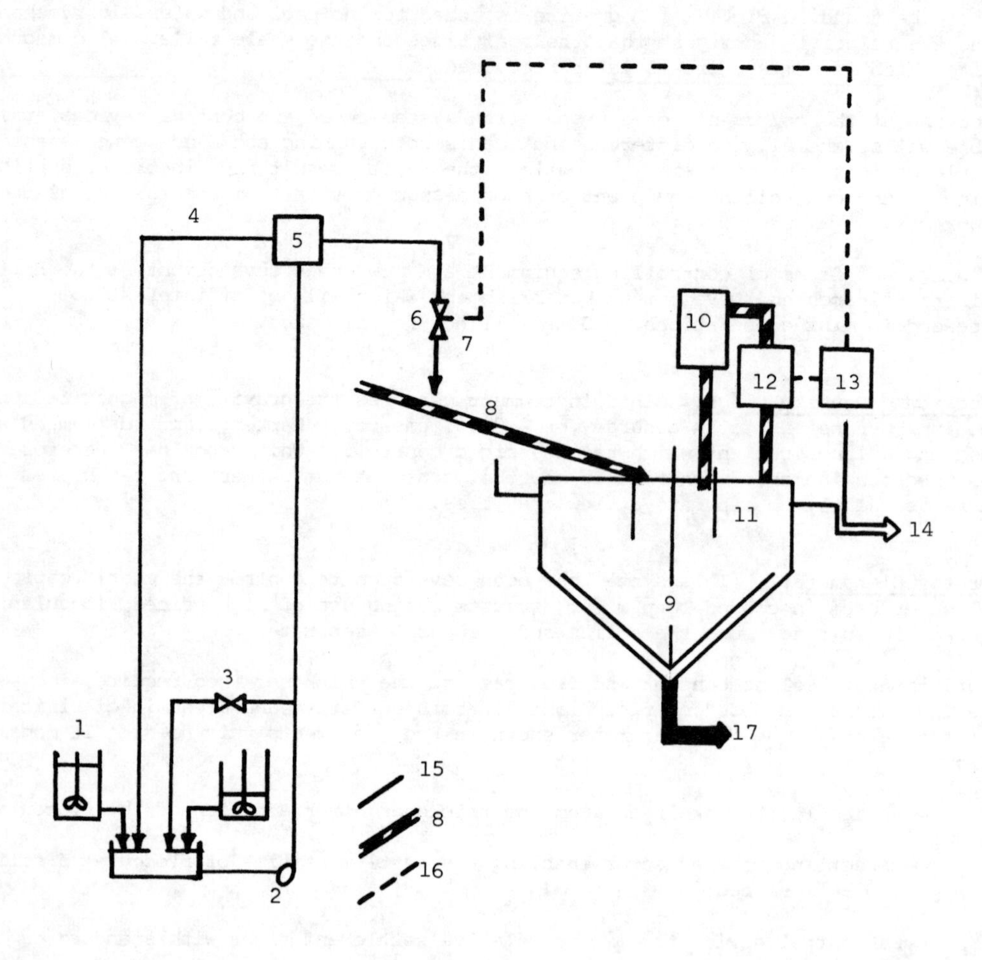

Fig. 26 NCB Clarometer

1. Flocculant mixing tank
2. Pump
3. By-pass valve
4. Overflow
5. Head tank
6. Control valve
7. Flocculant addition
8. Thickener feed
9. Thickener
10. Automatic sampler
11. Curtain
12. Settling rate detector
13. Control device
14. Clarified water
15. Flocculant
16. Control signal
17. Slurry

or back-scattered, depending upon the elemental composition of the coal. Elements of low atomic number, which constitute the combustible elements, back-scatter well and absorb relatively little of the back-scattered X-rays. Conversely, the report tells us, elements of high atomic number, which are the ash elements, back-scatter less and absorb relatively more of the radiation. The ash content can be determined from the back-scatter radiation.

Bretby Autex Flocculant Mixer. This device was developed about three years ago. The polyelectrolyte reagents are dispersed automatically to the necessary concentration in water. A number of units, over 30, have been installed and are working well showing a considerable saving. The system cost around £5000 in 1976 which is not expensive, bearing in mind the saving in manpower and flocculant which results.

The unit is shown in Fig.27. The Autex Unit - 15 - which disperses the polyelectrolyte flocculating reagents is made up of a flocculant inlet, water inlet, dispersing tube and an outlet for the dispersed flocculant.

Rotating screen. A rotating probability screen has been developed. This has a number of "radial spokes" mounted horizontally, which rotate and act as a screen, separating the coal into constituent sizes.

Preparation Plant Monitoring and Control

A number of simple complete automatic systems have been successful to the point that mini-computers and micro processors are now coming into use. It is anticipated that the use of central processors appear likely to result in reduced manpower, improved efficiency and better use of the plant. Use of a memory facility capable of recording the performance of different parts of the plant providing printouts or reports for management, as needed.

There are two situations to be considered. Completely new and computer controlled plants on the one hand and some degree of automation, as applied to older existing collieries.

Already the older mines have individual control panels installed. The addition of micro-processors would provide a number of advantages. These cover:

- quicker start up and shutdown
- longer washing time
- improved through put
- centralisation of control
- up to date information on operation and control of unsupervised processes

A further apparent advantage might be the elimination of human error, but since the input to all computers is through human endeavour, the nature and frequency of the errors only, will change. There will be less errors, but those which occur will be bigger, if not better.

The British National Coal Board through their Mining R & D Establishment have adapted a MINOS system for control of - coal clearance - control and monitoring at

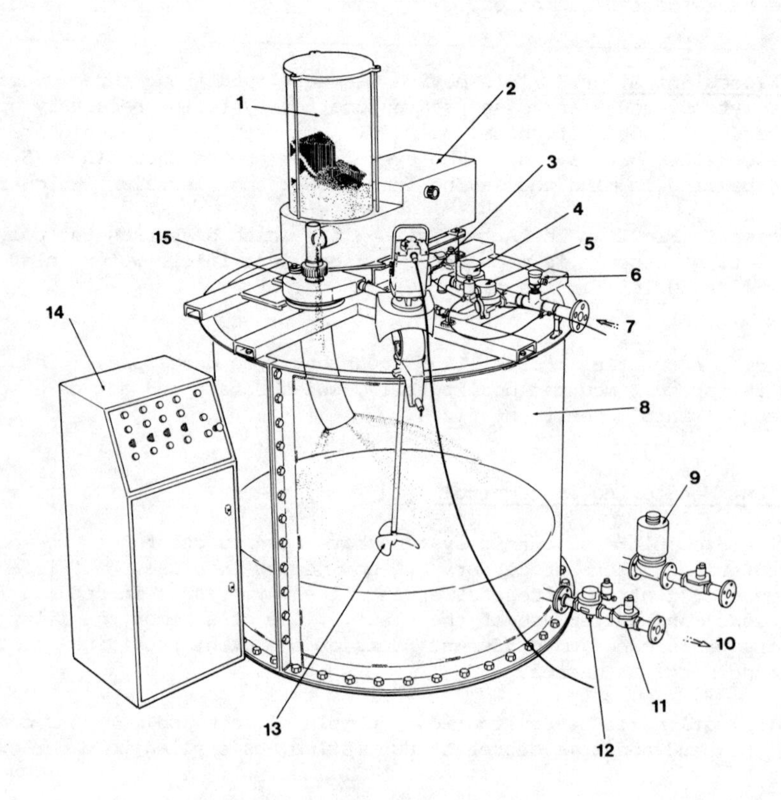

Fig. 2.7 Bretby automatic flocculant mixer

1. Powder storage
2. Powder feeder
3. Pressure switch
4. Mixing tank inlet valve (disc-type)
5. Autex inlet valve
6. Flow sensor
7. Water supply (clarified or mains, maximum pressure 207 kPa)
8. Mixing tank
9. Alternative arrangement (pump feed)
10. Mixed flocculant to stock tank
11. Flow sensor
12. Mixing tank outlet valve (gravity feed)
13. Impeller
14. Control console
15. Autex

collieries including environmental monitoring. The main features of the system are, according to NCB:

Monitoring. a) Visual displays which can be selected by the operator offer:

- mimic of plant network - brief status of each belt and bunker
- key conveyor delay summary
- faults currently present
- status of each bunker and belt
- conveyor delay log
- data transmission system status

b) Automatic alarm reporting:

- alarm messages in a meaningful way on visual display with associated three level audible alarm system
- display of variations from set norms

Control. a) Remote manual control:

- start stop conveyor motors scoop couplings, bunker hydraulic power packs, vibrators, metering all conveyors
- sequence start/sequence stop: sequence starting and reverse sequence stopping of conveyors to prevent overruns and chute blockages at transfer points
- load/discharge: remote bunker movement control
- set demand: to set bunker discharge rates
- regulate/unregulate: switching local closed-loop central circuits on bunker outfeeds
- lockout: to prevent inadvertent starting of plant, possibly during maintenance

b) Automatic control:

- automatic fault tripping. Optional/obligatory tripping on signals, local stop, local lockout, belt slip, motor overheating, belt alignment, bearing overheating, smoke brake alarm, belt sequence, water curtain
- simultaneous stopping of conveyors
- automatic restarting of conveyor sequences under non fault conditions and in response to specified bell signal sequences
- optional tripping of outbye conveyor in case of chute blockage

- staggered starting for shared transformers
- bunker management
- stone/coal segregation batching and bending on belts

c) Local Control:

- automatic monitoring
- provision for surface lockout

Plant definition. Automatic question/answer on plant and data transmission system characteristics.

Maintenance and testing. Fault finding, testing and maintenance.

Management information. Automatically end-of-shift summaries.

Preparation already plays an extremely important part in coal production. Standards will grow no less while moisture, ash and sulphur will all tend to rise as dust suppression and hydraulic transport, removal of greater quantities of roof coal and less desirable areas throughout the world are mined.

FLUIDISED COMBUSTION

Indirect reference has been made to the consumer's needs, particularly with reference to controlling moisture, sulphur and ash. However a method has been developed which enables fuel, irrespective of its ash or sulphur contents, to be burned efficiently at high combustion intensities, while keeping the emission of sulphur dioxide and nitrogen oxides well below any of the standards already set,as yet to be proposed,and which in some parts of the world are enforced rigidly.

The relatively low combustion temperature and the environment in which combustion takes place, reduces to a minimum the formation of corrosive substances likely to attack metal surfaces in conventional plants. The temperature of the bed is maintained by working fluid passing through the heat-transfer tubes located within the bed. This, coupled with a much reduced heat-transfer area result in overall reduction in plant size plus significant savings in costs. It enables low quality fuels to be burnt efficiently and clearly in a way previously impossible in conventional furnaces.

Much of the research and development on fluidised combustion was carried out in Britain, beginning at the Central Electricity Generating Board's Marchwood Laboratories. Later the NCB took over the project at its Coal Research Establishment and BCURA Laboratories.

In Autumn 1974 the International Energy Agency (IEA) was founded, made up of member countries of OECD. One of the main objectives of IEA is to produce alternative energy supplies by promoting research into new sources of energy and developing them. Coal was one of the Working Groups set up in which Britain was invited to play a leading role. Mr. Leslie Grainger, at that time Member for Science to the NCB, was invited to become chairman. The fluidised bed project, one of five proposed, was supported by Germany and the U.S.A. in addition to the

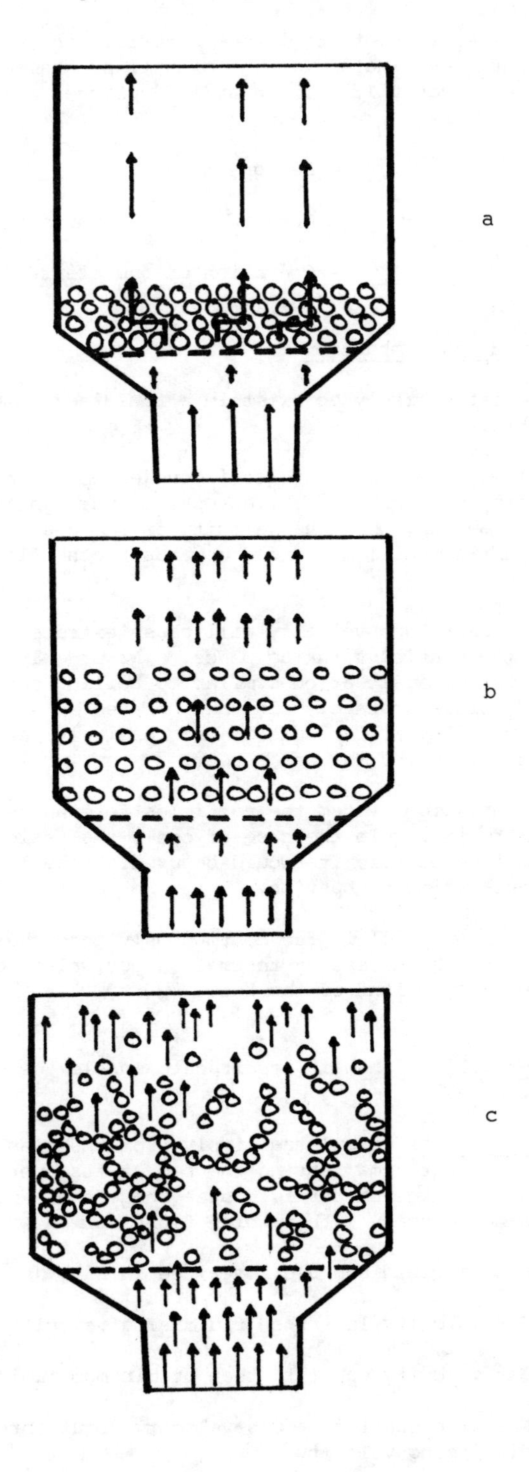

Fig. 28 Fluidised Combustion

United Kingdom, as equal partners in the construction of a plant at Grimethorpe Colliery, near Barnsley, England. The overall programme is scheduled to last seven years and will cost £13 million at 1976 prices. It was divided into three phases:

- Design Study
- Construction & Commissioning
- Operation of the plant

Principle of Fluidised Combustion

There are three states likely to exist in a fluidised combustor, which are illustrated in Fig. 28.

When a bed of finely divided particles is subjected to an evenly distributed upward low velocity flow of gas, the gas passes through the bed without disturbing the particles. However, when the velocity of the gas increases, the time will come when each particle will be forced upwards by the flow, becoming suspended in the gas stream.

Any further increase in gas velocity will cause extreme turbulence in the bed with rapid mixing of the particles taking place. No orderly arrangement of the particles persists and a situation similar to boiling occurs. The bed is then said to be in a fluidised state. It then behaves as a liquid. A further increase in gas velocity will result in progressively larger particles becoming entrained in and removed by the gas stream.

In a fluidised combustion system the particles forming the bed are made up of suitable inert material. In the case of coal, the residual ash from earlier coal combustion, limestone or dolomite would be used. The limestone is added to absorb the sulphur during combustion.

Solid or liquid gaseous fuel is fed continuously into the bed in the quantity required to maintain the necessary thermal output, although the quantity of carbon rarely exceeds 0.5% by weight of the bed.

The fluidised gas will be the air required to enable the fuel to burn.

The fluidised velocity is calculated simply from the plan area of the bed and the volume of gases per unit time leaving the bed at its operating temperature. For a particular bed with a given particle size range there will be an upper and lower limit of fluidising velocity within which fluidisation will take place.

The three states of a fluidised bed contained in Fig. 28 are:

a) Gas velocity less than fluidising velocity

b) Gas velocity equal to that of minimum fluidising velocity

c) Normal operation. Gas velocity about three times minimum fluidising velocity

There are a number of potential applications for the fluidised bed:

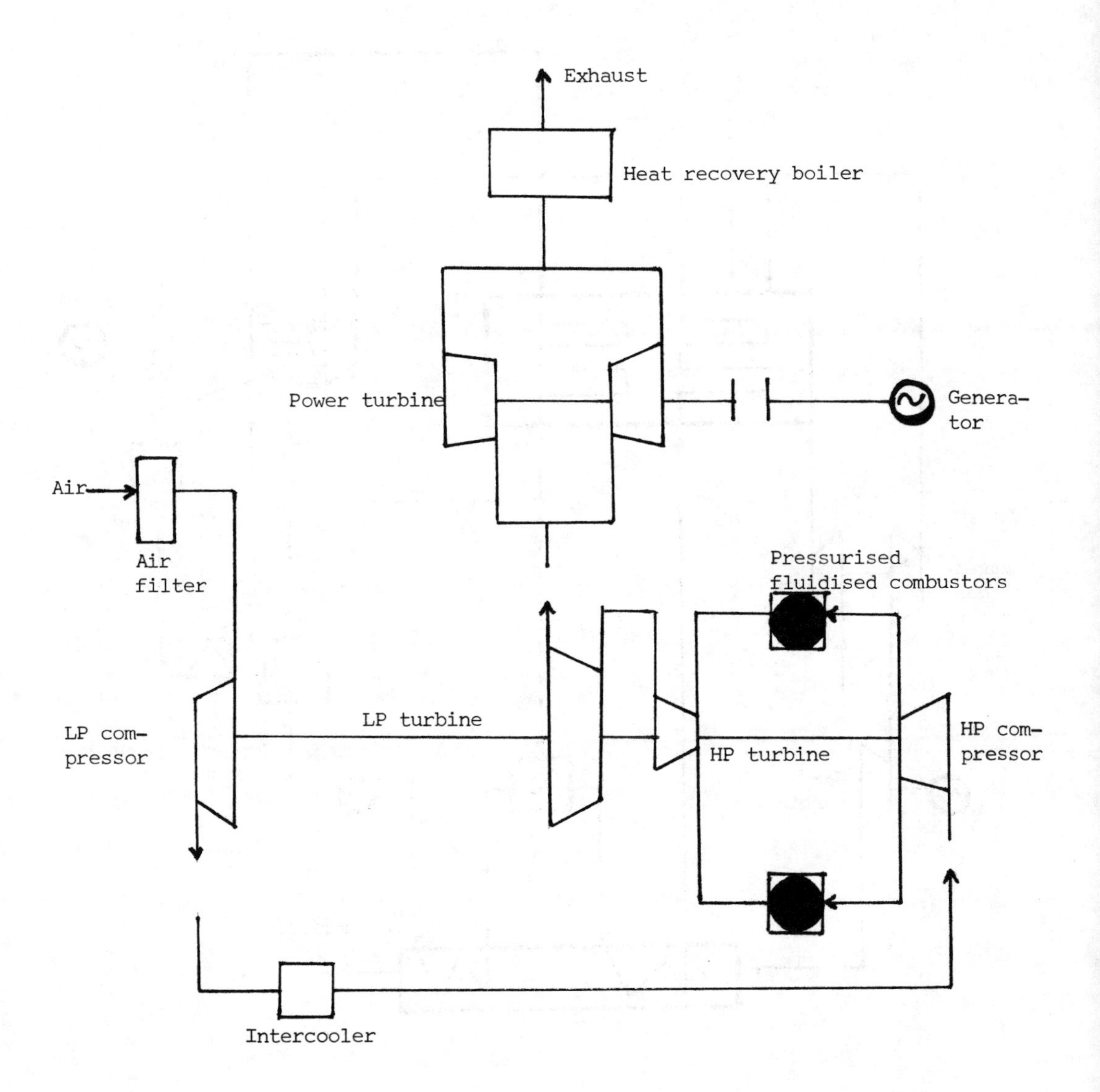

Fig. 29 Pressurised fluidised combustion air heater plant flow diagram

- Steam raising pressurised fluidised combustor and gas turbine plant illustrated in Fig. 30.
- Pressurised fluidised combusion air heater plants contained in Fig. 29.
- Fluidised combustion for sludge incineration

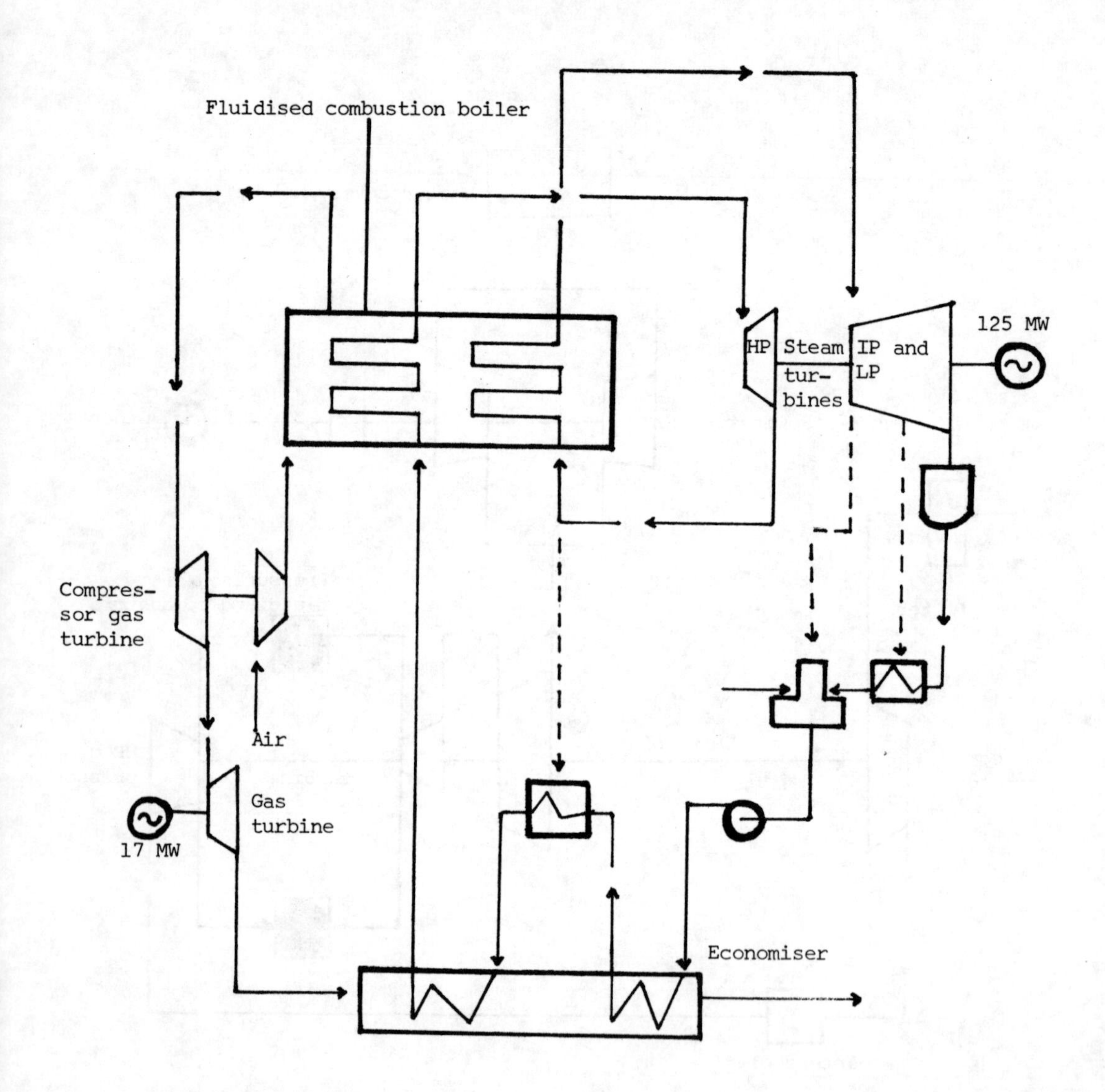

Fig. 30 Pressurised fluidised combustion steam generation flow diagram

Clearly the use of a fluidised bed, whether for steam raising or gasification - as will be discussed in chapter 9 - with a capability for removing sulphur will make the task of coal preparation that much less difficult.

Chapter 8

THE MINE OF THE FUTURE

The mining countries of the World are striving to produce coal using less men, underground just as for the past fifteen years progress in mechanisation has removed some of the physical effort and danger from the winning of coal. But as one problem seems to be solved so another, dust, reared its head. The nature of the problem is continually changing.

But in deciding upon the nature of the Mine of the Future it is vital to consider how the product is to be used. Legislation will clearly be a limiting factor, because it is obvious that people will continue to demand an environment which is not polluted through emission to the atmosphere.

Controls in the States, Germany and Britain are already tough and likely to stay that way. Producers must therefore ensure that the coal is available in such a form that it can be used as produced or in a modified form. Alternatively the means of burning the coal and dealing with any pollutants which might result in the gas emitted must be available.

Most of the mining nations as was seen in Chapter 4 are seeking means of achieving this end through their research programmes.

However, it should not be forgotten that the gap between research and ultimate development into a reliable piece of equipment or technological method, probably equates to a time scale of about ten years.

One of the main objectives underground must surely be a co-ordinated system containing the elements of an automated long wall face, to include:

automatic shearers

supports

conveyors

face end equipment

Clearly the components have been developed, but if the successor to this arrangement is to be available during the first decade of the next century, work must be put in hand without delay. Then general adoption should be possible some ten years later.

Forecasting the progress of technology always presents a problem as a British octogenarian reminded us recently, his mother who died young had never heard a movie, seen television or a man put on the moon. And yet all these things had happened during her sons lifetime.

One probable source of innovation will come from the spin off from other technologies which may be applied to coal production. Space travel resulting in a whole range of new thinking and new developments is a good example.

But this can only be speculation and therefore the main areas of mining activity should be reviewed. The detail is to be found in preceding chapters.

Coal Reserves

The reserves of coal really amount to that believed by experts to be capable of recovery; to which should be added that detected by the geologists. But certain features need to be considered, in a national context:

- The nature and characteristics of the reserves.
- The way in which the reserves will change by the turn of the centry.
- The allocation of reserves likely to be recovered by the turn of the century and the factors which may limit their recovery namely:

  Roof coal

  An increased area of extractable coal

REMOVAL OF COAL BY MECHANICAL MEANS

There are a number of alternatives to be considered in this field:

Mechanical Cutting

Use of Explosives

Nuclear Explosives

Hydraulic Mining

Use of Lasers

Drilling

Mole and Telechiric Mining

With these needs in mind the changes to be expected in the years ahead will be examined bearing in mind the currect practices described in Chapter 3.

Mechanical Cutting

Tools are required and used to reduce the coal seam to fragments of a handleable size. These tools contain picks, clusters of picks or cutting heads attached to machines. One important factor is the power requirement since efficiency in cutting is often expressed in these terms.

However, dust produced and the degree of methane are also factors which are particularly important.

It is generally claimed that mechanical shearing at its most effective is more efficient and superior to hydraulic mining and other methods. Tough materials vital to a shearing capability must be improved. Success here would provide an added incentive to go mechanical.

Today, Tungsten Carbide in a cobalt matrix is standard for cutting coal. It is not used for rock because when used in that role it chips too easily.

Other materials such as Nitrides and Berrides have been proposed in view of their hardness, but unfortunately they have proved to be too brittle. Tungsten, although possessing a good level of abrasion resistance including resistance to chipping, is not ideal.

Other means have also been suggested, including surface hardening by Laser beam and heat treatment. Other alternatives have been examined too, including cutting with discs, but the experts have viewed this idea without enthusiasm. Considerable improvement in cutting techniques is to be expected in due course.

Explosives

The use of explosives is the only reliable method of cutting hard rock, underground. In terms of mechanical efficiency explosives are the equal of mechanical cutting, but the method is labour intensive, it cannot be mechanised easily and considerable hazzards are involved.

Recent developments have included slurries containing explosives as opposed to nitro-glycerine. These slurries are extremely safe but require a strong detonation.

A number of explosives of greater versatility have been tried. Also methods, for application to mining, as are used in the cutting of girders, jettisoning of canopies in aircraft and linear chord cutting have been considered. But all have defects demanding that alternatives be found.

Nuclear Explosives

There is considerable potential in the use of nuclear explosion (PNE), but the process is limited to areas where the explosion will be contained and so the result is an enclosed hole. The coal in the area would be broken up and so assisting in-situ conversion. Radio active contamination will often result however as well as damage to structures on the surface, the former denying men access to the area. Its use must be restricted.

Hydraulic Cutting

This method has been described earlier under hydraulic mining, applicable to soft coals and those such as in Germany which are not easily accessible, or form steep seams. It has been practised in many countries - with particular success in some. In most instances pressures have been less than 5000 psi. Two sizes of jet are likely to be used; one of a large cross section for the hydraulic movement of coal to take it away from the face. The quantities of water involved are very considerable and are capable of causing deterioration of the roof and floor. It is also vital that the gradient of the seam enables the water to flow away quickly. The arrangements within the mine must ensure that this happens.

Much higher pressures have been used, the Mining Research and Development Establishment possessing facilities capable of achieving 200,000 psi with corresponding high water volocity and beam like jets. The impact is capable of producing narrow slots in hard materials such as rock, but power consumption is very high - about ten times that for mechanical cutting. This is a disadvantage.

However, if the jets are pulsated instead of being trained in a continuous jet, large slabs of rock are removed from the face.

A further possibility is to include jets in a machine but the problems of using such moving mechanisms and very high hydraulic pressures, coupled with the fragile nature of jet nozzles, are considerable.

There is little doubt - a view confirmed by those involved in the experiments - that pulsated jet cutting together with mechanical methods of breaking may produce an efficient means of removing coal or rock. But such a process is difficult to control and as a result lacks accuracy.

Rock cutting is therefore the most likely application for this combination particularly in drivage situations where roof supports should fit tightly. Again quantities of water and use of power are bound to be high.

There is clearly a very long time scale which must elapse before a demonstration unit is available.

Drilling

One of the methods still to be discussed in detail - in-situ - will depend upon accurate drilling both for underground gasification or solvent extraction, or for acess to holes blown by nuclear explosion. It will also be needed to aid the removal of methane in conventional mining.

Most of the drilling expertise does in fact lie with the large oil companies from their very considerable experience of drilling in the United States, Middle East and more recently in the North Sea. Considerable scope lies in transferring some of the principles to the mining industry, but this will clearly take time.

Subterranean Mining

As has already been indicated on a number of occasions methods of mining coal which would not otherwise be accessible need to be developed further and perfected. These situations include thin seams, added to which must be the very considerable reserves which lie under the sea.

Already conventional mines project for some miles under the North Sea, but clearly long lines of communication become extremely expensive. Some method of removal by remote control needs to be developed. Clearly vertical drilling is one such possibility accompanied by gasification or liquefaction, both of which will be discussed fully later in this study.

Another approach called Mole Mining has been proposed by a British Professor at Queen Mary College London.

This involves the vision of a mechanical mole burrowing underground, presumably in extension of the in-roads already made under the sea. This mole would consume coal which would then be discharged to the surface by hydraulic means through a series of pipes. Similarly, the air and fuel would be transmitted in the opposite direction to operate the machine, while co-axial cables would take the signals to control and steer it.

The coal containing slurry is delivered direct to the point of use and burnt without the usual preparation.

Some researchers have pointed out that the main driving force would have to be enthusiasm, a force which would be unlikely to sustain the roof or enable the cable involved to be moved beyond a certain distance from base.

However, even the most sceptical accept that telechiric technology is worthy of examination.

IN-SITU EXTRACTION

The Coal Mine of the future will be dealing with coal reserves comprised of seams:

- at present being worked by normal mechanical methods while deeper seams must be sought as last deposits are recovered.
- below 4000 ft. deep

The first group will present increasing problems and become more expensive with time and depth, while those below 4000 ft. deep will have the additional problem of high strata pressures and temperatures.

It is when both groups reach a point of uneconomic and impossible conditions for mining, that in-situ methods may be the only practical means of extraction.

The principles of In-Situ Extraction Processes have been discussed in Chapter 9 with regard to underground gasification, which may be summed up in these terms:

Method of Extraction	- Gasification
Method of Transport	- Hot air and gases
Operating Temperature	- Up to 1100°C
Operating Pressures	- Up to 50 bar
Product	- Low Btu gas
Yield	- 60 to 70% available coal
Benefits	- reasonably developed techniques
Drawbacks	- Gas - low calorific value Process - high temperature

Other methods offer alternative means of recovery. They include:

Pyrolysis

In this method complete gasification is not attempted, the volatiles being driven off, leaving a coke residue. Although the residue of tars deposited in cracks and fissures may tend to block movement of the gases, attempts have been made in the U.S. to produce higher hydro-carbon - with inconclusive results. Using the previous method of report for easy comparison, the data is;

Method of Extraction	- Heating
Method of Transport	- Hot gas
Operating Temperatures	- 800 to 900°C
Operating Pressures	- Low
Product	- Gases and Tars
Yield	- 20%
Benefits	- Coal liquids and char capable of gasification — produced
Drawbacks	- Blockages from tar deposits

Quenched Combustion

Certain Oil Companies involved with coal, such as Shell and BP, employ a technique using oil in which oil is fed into the hole with air and burnt. Water is then pumped in to follow the flame front and produce steam. This could offer a possible application for coal. Details are:

Method of Extraction	- Combustion
Method of Transport	- Steam
Operating Temperature	- 800°C
Operating Pressure	- Low
Product	- High Temperature Steam
Yield	- No information
Benefits	- Low temperatures
Drawbacks	- Water additions difficult to control

Total Combustion

This form of gasification involves completion of the combustion process, producing inert gases from which the sensible heat can be extracted. Experiments in the U.S. are proceeding with the idea of recovering the energy from abandoned pillars.

Method of Extraction - By combustion

Method of Transport	- Using Hot Gases
Operating Temperatures	- 800 - 900°C
Operating pressure	- Low
Product	- Sundry - Hot
Yield	- 90%
Benefits	- Easy to operate
Drawbacks	- Exothermic reaction

Solvent Digestion

Solvent Recovery is being examined and developed by a number of countries - USA, UK and Europe. They all involve the digestion of coal using oil based slurries, usually anthracene oil. Some processes being examined involve hydrogen transfer.

Method of Extraction	- Solvent Digestion
Method of Transport	- Hot Solvent
Operating Temperature	- 450°C
Operating Pressure	- Up to 150 bar
Product	- Coal Slurry
Benefits	- High yield of liquid product
Drawbacks	- Solvent is expensive - high solvent coal ratio. Liquid product highly viscous.

Aqueous Phase Liquefaction

Water at high temperature together with synthesis gas is in theory, circulated through the coal seam.

Method of Extraction	- Liquefaction
Method of Transport	- Gas and Water
Operating Temperatures	- 180°C at 10 bar
	- 250°C at 40 bar
Yield	- about 50%
Benefits	- Fairly low temperature

Superficial Gas Extraction

The use of a superficial gas is made in a pyrolysis process to remove the volatile matter. It is not visualised as an "in-situ development.

Method of Extraction	- Solution is a super-critical solvent
Method of Transport	- Super-critical gas
Operating Temperature	- 450°C
Operating Pressure	- Up to 100 bar
Yield	- about 30%
Product	- Liquid
Benefits	- High seam penetration through low viscosity. Solvent recovery by depressurisation.
Drawbacks	- Poor yield
	- Hazardous Solvents
	- High Operating Pressure

There are three other methods which should be mentioned, one of which, Hydraulic Mining, has been reviewed very fully. The advantages include low temperatures and simple in operation. On the other hand the range of operation is limited.

Drawbacks in In-Situ Processing

A key factor is the surface to volume ratio. This may be achieved through

- Horizontal boreholes
- Hydraulic fracture
- High explosives producing fracturing ratios including 600 tons of coal per ton of explosive. Development of slurry explosives mentioned earlier by ICI is a useful advance.
- Nuclear explosives which have also been described.

Extraction media. These are usually introduced by pumps from the surface. Heat necessary for underground processes at high temperatures is usually added with the extraction media. Heat is also vital for the endothermic reactions.

Recovery of Extraction Products. Sufficient pressure must be available to force the products to the surface. An aid such as the addition of gas or water may be needed.

Recovery of Extraction Media. The amount of media required to fill the space left by the extraction process and the removal of the product will increase with time. Some means of stabilising the size of the void must be found, unless unreasonable amounts of extraction media are to be used. This could be achieved through:

- addition of water or rubble to displace the extraction media and product.
- materials which vaporise, leaving behind a low density vapour.
- caving of the roof under controlled conditions.

Product Preparation. Hydrogen sulphide and removal of solids may be required. Undesolved solids and ash will also require removal before recovery of the solvent for recycle.

A number of technical problems will need to be solved most of which have been discussed already. They are worth repeating:

- Boreholes must be drilled accurately of sizes varying from those needed for access to the seam, to large diameter, up to 60 cm. for introduction of explosives. While the expertise is available within the oil companies the process is slow and the costs very high.
- Fracturing of the Coal Seam has been seen to bring benefits although in order to prevent by-passing or blockages with various liquids an even permeability is essential. The high pressures persisting at increased depths makes these difficulties greater.
- Considerable surface construction for solvent digestion would be necessary, while the quantities of extraction media required would place a considerable strain upon the production of anthracene oil.

Costs. The factors which affect the cost of the product have been discussed fully in Chapter 9 as applied to gasification.

Shell Coal International have also carried out an exercise regarding cost using a process based upon a coal solution, but details are not to hand.

Conclusions.

a) The constraints on seam thickness and host rock make most methods suitable for a very limited number of seams.

b) Environmental problems are associated with most processes.

c) There are considerable technical problems which require much more research.

FUNDAMENTALS OF MINING

Most of the operations connected with mining have now been virtually mechanised, but the time for which the machinery is running is around 35% only of that available. Machines unfortunately are not sufficiently reliable particularly the armoured face conveyor and the power loader. Roof control leaves much to be desired as a result of interaction with other workings and even failure of powered supports. Co-ordination between installed face capacities and shaft capacities is not always good. Also, the rate of drivage is not very predictable while faults seem to catch out the best of mine plans. However, means of forecasting problems ahead are now available as discussed in Chapter 5.

The means of recovering coal have in fact advanced at a faster rate than their application and control and so all this progress has been achieved at a lower percentage of the seam recovered. The figure is in fact half that of the immediate post war years and it has been estimated in Britain that roof coal is left on some 60% of coal faces as a result of abandonment due to unexpected faults. Similarly because seams under 2.5 ft. cannot be mechanised and are therefore immensely expensive to work, total reserves are being reduced continually. In addition pillars are left to protect roadways resulting in a further loss.

Mining Needs.

Two requirements stand out. Recovery of reserves must be increased so that the level returns to the post war figure while machines must be made more reliable. The lessons of continuous working examined in Chapter 4 must be taken to heart.

But there are fortunately pointers which bring hope of this being achieved.

- Methods of seismic detection as practised at Selsby in charting the new mine there will arm managers of other mines with warnings of faults and the pattern of faulting in a way not used universally at present.

- The need to leave pillars to protect roadways and roof coal to maintain weak roofs will be supplanted by shield supports and monolithic packing now being practised in the most advanced mines.

- Hard rock tunnelling machines incorporating support systems for tunnels are fast becoming available particularly from German sources.

- New systems of transport including hydraulic pipes and rubber tyred vehicles are available and in use. In turn reinforced road surface systems are being developed so that roads can be used twice. There is reference to this work in British research reports.

- Shaft capacity is being increased but full benefit from matching face capacity with that of the shaft, may have to wait for the development of new collieries and mines, or at the very least extensive modernisation plans.

- Reliability and measurement through mining systems remotely or automatically controlled is on the way. However it must not be overlooked that complete automation, removes from the scene, men who would otherwise make good shortcomings in individual components of the system. Then there is the likelihood of the whole arrangement grinding to a halt on occasions.

But perhaps the most important factor is the incentive for men to make the new systems work. There is little doubt that in many European countries where labour is expensive and there is a lack of inclination to go into mining, the amount of coal required to help fill the energy needs will rise. Automation will not mean a loss of jobs, long term. The age range pattern shown in Table confirms that there is a bulge at the top end. Any reduction in the age of retirement will easily account for those underground who may not be needed.

However, by linking coal brought to the surface with payment, a definite incentive will be offered for maximum coal possible to be removed from the seams. In turn this means that the maintenance men will ensure that they play their part so that machines keep going; they will be part of the team, they will share in the benefits of payment by results.

Incentive bonuses introduced in Britain in 1978 by the National Coal Board has had a dramatic effect upon production which prior to that move had been falling disastrously. At present the quantities of coal being brought to the surface are beginning to create an embarassing situation, in markets, although not as part of the strategy, long term.

The Realities of To-morrow.

Having looked at the available technology, including the improbable, both in practice and under development, it should now be possible to picture the sort of mine which might be in existence at the turn of the century. But it must not be forgotten that markets are very small and development costs extremely high. While the technology may be available, it could be totally uneconomic to employ it. This probably applies to an even greater degree in conditions which point to increased unemployment as being a fairly permanent feature of our way of life. Unless means are found for sharing work available then it may be that the unemployed will demand a greater degree of labour intensity in our general industrial activities.

It must also be remembered that much of the mining in years ahead will take place in mines which are already established and have many seams still to be worked.

However, in general terms for optimum co-ordination of processes to occur, mines must be large - between 3 and 5 million tons production a year. Continental mines began this pattern, followed by the United States, and now Britain is joining in. Using the Selsby (UK) experience it is clear that the means are available to organise the coal faces to obtain the best conditions within a known pattern of faults and disturbances.

Two methods of recovery obviously come to mind:

- Longwall, producing some 1 million tons per year.

- Shortwall, taking out pillars, with high efficiency and introducing a flexibility to the general planning of the mine.

The real key must always be management, making use of the confirmation obtained from extraction of fault patterns to obtain the best use of men and equipment as well as the recovery of the reserves.

Coal faces. Coal is coal, and the formations which have taken millions of years to develop will not change over night even though the equipment will change. We can expect the components already examined and described, to form a total system. Equipment and components will also be standardised. The shearer without doubt will still be operating stripped of its chain drive, with picks of improved hardness although with gear ratios such that dust is not encouraged. The attachment of sensors to replace two picks in the drum, which react to cutting forces, will ensure that more coal is removed. Alternatively a sensor placed behind the shearer will react to the natural gamma rays relaying a signal back to a controller or computer which will achieve the same end. The supports at least with respect to advance mining will probably be in the form illustrated in Chapter 6 in the campacker, which is at an advanced state of development. Otherwise supports will contain both a high setting load and adequate resistance to lateral movement.

Shortwall faces will be used to extract those pillars of coal left to meet the needs of large longwall faces and also to provide flexibility where the mine plan requires run down or start up periods for face lines. The shortwall face will have simple deep buttrock systems, abandoning the roadways as they retreat. The NCB/Dosco In-Seam Miner represents the type of equipment which we shall most probably be seeing.

It must be made easily transportable from site to site.

The direction of mining will depend upon the incidence of faulting. Where conditions are good faces will generally be of the retreating type. Where this is not so advance faces will be used including shorter runs and a necessary reduction in the development drivage for each production unit.

The advantage of retreat mining is that it involves less men. It will therefore be used whenever possible in the interests of economy.

Transport

Without doubt hydraulic means will be employed for coking type coals while the belt conveyor will continue to carry fuel to be used in power stations, emerging via the shaft which carries the return air supply. Supplies from the different faces need to be synchronised and planned so that the flow of coal is as continuous as possible.

Local shafts will be a feature of new mines as in the U.K. mine at Selsby, but first a cheap method of sinking. Small shafts large enough for ventilation, water and fuel is needed.

If man riding could be included too, so much the better to eliminate the long walks to and from the coal face as in many existing mines. In the "mine of tomorrow" every man must be able to reach his point of working with speed and in reasonable comfort, as far as possible under conditions similar to those engaged by surface workers.

It must also be remembered that in addition to men and coal, transport is required for track laying and withdrawal machinery, a vital element in prospecting. All require some form of mechanisation, with the whole transport system being monitored constantly, but the real issue lies in the merit of free steered vehicles with tyres versus rail transport. Free steered vehicles if automated would require remote steering while the problem of passing would have to be mastered and controlled. Twenty years ahead is difficult to forecast in this field because much work is already proceeding on a number of novel forms of transport for use in mines. These include reduced pollution combustion engines, improved electrical storage cells and liquid nitrogen as a source of fuel.

Trolley-wire as a method of electricity supply is one which may well be seen, provided present restrictions as to its use are dropped. It could be employed in conjunction with storage batteries or possibly liquid nitrogen. The auxiliary powere needed could be used where there were no conductors, and traction over short distances is needed; also as a means of power for safety devices. This would enable the external source of supply to be cut off at times of potential hazzard but at the same time permit limited operation to continue elsewhere.

Roadways will also play a prominent part. Paved roadways will be very evident to carry rubber-tyred vehicles, jeeps for small groups and bus-like vehicles for larger teams, with slightly larger vehicles or trains for movement of materials.

This will require an understanding of rock mechanics with rocks protected from pressures either naturally or artifically. Layout will play a very important part.

Roadway Drivage

We can expect drivages to be fully mechanised so that the tunnel will be cut to preplanned design and supports introduced as the device advances. It is expected that short boom cutter types, already developed, are more likely to be used for stone cutting, rather than the full face type.

Coal Preparation

It must be assumed that a coal preparation plant will be highly automated, of modern concept and capable of 2000 tons per hour throughput.

Many of the existing preparation techniques will still be in use, but others will be developed to offset some of the properties of coal which create difficulties and which even modern techniques are unlikely to reduce.

- Improved mechanisation is likely to ensure an increasing proportion of fines.
- Hydraulic transport will probably increase the moisture content.
- Higher coal face recovery will raise the amount of dirt and shale required to be removed.
- Sulphur content could well increase although in many parts of the world, the States and Germany, it creates considerable problems now.

It is here that considerable effort must be made so that the final product reaches specification without the use of labour - a long haul from the days when men and women sorted the coal by hand.

One of the problems clearly is to perfect the means of obtaining a signal capable of transmitting an accurate picture as to moisture content and ash present in a coal. Until this is achieved the advances which we expect to see in the mine of the future will remain in the days ahead.

Clearly computer controlled automation will be the main feature but in parallel, there will be a need for trained supervisors of high calibre to exploit and control this highly technical field.

Remote Control and Monitoring

If the object of progress is to remove as many men as possible from underground then clearly overall control will be established on the surface.

Each of these control centres will be interconnected and also in touch with all aspects of mining - working groups and transport.

Closed circuit television will be a common occurence just as it is used to monitor operations in sewage works, or to watch the activities of shoplifters. Activity in the mine will be reported at regular intervals and bottlenecks spotted quickly. Hold ups mean lost production and lost return on Capital.

There is clearly much to be gained from the experience of the Post Office, Communication Companies and the Armed Forces, all of whom share the same problem of keeping in touch with men and processes. The development of small light portable personal equipment is vital.

All this will lead to a far better co-ordinated operation and in turn greatly increased productivity.

Forecasts as to the likely improvements can be made from the review of mining activity contained in chapter 3. A recent prediction by the Director of the British Mining Research and Development Establishment included these suggestions:

Longwall - 5,000 tons per day

Shortwall - 1,500 tons per day

For a three million tons per year colliery there will be two longwalls, three shortwalls, with a spare set of equipment for each.

Twenty men will be required for a longwall face with ten to a shortwall - over 24 hours.

Transport will be simple but fast, with fifty men required. Tunnel and face development groups will have five main drivages needing seventy five men.

If a further hundred people are required for management the mine of tomorrow will employ 550 people producing 5,500 tons per man year against the present twelve hundred needed in a British mine today.

It would be interesting to learn how this view compares with targets set by other mining nations, bearing in mind the low starting point in British mining productivity during the mid-seventies.

Environmental Control

Any reference to Monitoring must swiftly lead to environmental control. As seams become deeper, so the cost of dealing with high temperatures will rise. The cost of refrigeration applied universally to a mine will be excessive. Air cooled suits may well be the solution with air conditioning for man riding vehicles. Monitoring of conditions will be as important as control of the processes, because if a reduction in the level of manpower visualised comes about there will be few men available to deal with potential dangers. Today's methods of monitoring involving the tube-bundle system and fixed point sensors placed in airways must be updated to include microprocessors, improved data transmission and computer systems. More is required than fixed point monitoring and it is quite probable that an infra-red laser tuned to the absorption frequency for methane could be used to detect the gas over a wide area and report the fact back through the computer to keep management informed.

Future Trends

Coal will, without doubt, be used as a feedstock for the chemical industry and petrochemical processes as well as synthetic fuel production. All these processes involve coal conversion, which must include:

- gasification, particularly medium Btu and SNG.

- liquefaction to produce liquid hydrocarbons.

- pyrolysis to obtain oil and gas from coal.

Whichever way, the coal must first be mined, but the manner in which this will be achieved will vary in different parts of the world, according to hardness, ease of access and strata encountered.

The most likely method of mining will be by conventional means backed up by "hydraulic". But it will be a very long time before all men work above ground. One is tempted to say if ever.

It is likely, therefore, that the three vital processes listed for the conversion of coal will be carried out above ground until at least well into the twenty first century. Even the American programme, with its vast research funding does not anticipate the development of liquefaction before the year 2000.

Automation covering recovery, transport preparation and control will move forward apace while conditions underground will continue to improve, particularly with regard to personal comfort.

The Linear Motor Train illustrated here is a good example of one of the contributions to comfort.

GMT Linear Motor Train

This vehicle is at present under test at Bretby in preparation for the Mine of Tomorrow. Whatever technology may offer, in the ultimate it will be the men on the job who will be needed to operate it successfully.

Chapter 9

GASIFICATION OF COAL

No-one can be in any doubt as to the problems which will emerge as oil and natural gas first fall short of demand, and later are phased out as reserves which are easily accessible dwindle.

Capital will have been spent upon the transmission and distribution gas systems now carrying the Natural Gas Supplies not only from the North Sea to Britain and Northern Europe but also from as far away as U.S.S.R., Iran and North Africa, to France, Germany, Italy and Switzerland. There is a considerable dependence upon gas, although perhaps in a different way to the needs in the United States, where gas turbines are a feature of electricity generation. Certainly today, Natural Gas should not be used for steam raising and heating. Its virtues are far too numerous and it is far too valuable as a feedstock for the Chemical Industry - to be discussed fully later.

The most versatile of them all is coal. Coal can be burnt directly with considerable efficiency. It can be gasified to produce a low Btu gas or subsequently modified to Synthetic Natural Gas. Alternatively, the low Btu gas can be improved through removal of the Nitrogen to a medium Btu gas, believed by many to be the most suitable and economic form as a feedstock to the Chemical Industry. Coal can also be liquefied and later fractionated to produce the equivalent of petroleum, a range of aromatics and on to tars. Liquefaction is the only route from coal to a liquid fuel for transport. But production on an economic scale is a long way off - possibly the other side of the year 2000.

Without doubt, conversion is the Role of Coal in Tomorrow's Technology.

<u>Gas from Coal</u>

The essential reactions which occur in a gasifier during the process of making gas from coal are:

$$C + H_2O \;\_\_\_\_\; CO + H_2$$

$$C + H_2 \;\_\_\_\_\; CO_2$$

$$CO_2 + C \;\_\_\_\_\; 2CO$$

$$CO + \frac{1}{2}O_2 \longrightarrow CO_2$$

$$CO + H_2O \longrightarrow CO_2 + H_2$$

$$3H_2 + CO \longrightarrow CH_4 + H_2O$$

The product gas will consist of a mixture of CO, H_2, CH_4 and CO_2 plus impurities. The latter will be mainly N_2, H_2S and COs. In the case of fixed bed gasifiers, tar, phenols, higher Hydrocarbons and NH_3 are also produced.

There are three and only three proven large scale coal gasification processes.

- Lurgi - essentially fixed bed
- Koppers - entrained bed
- Winkler - fluidised bed

Looking a little closer at the processes, this view is seen to be growing:

Fixed Bed

This method has problems in that it is sensitive to caking and ash differences. It is unlikely to be adequate for the needs of the year 2000.

Entrained Bed

There is much in favour of this gasifier in that very fine coals can be used with oxygen being blown-in. The temperature tends to rise to a high level, resulting in ash and molten slag. Heat recovery is difficult, while the presence of Sodium and Potassium in the gas corrodes turbine blades.

Fluidised Bed

This type runs at atmospheric pressure. As can be seen from the qualifications accompanying each type, no single process is best under all conditions. Each has its particular advantage and disadvantage under specific conditions. All are good reliable means of producing gas from coal. All are of German origin.

Typical analysis for the three proven processes, listed in Table 21 , are - taken on a sulphur and CO_2 free basis.

TABLE 21 Typical Gas Analysis for three Gasification Processes

	Lurgi	Koppers	Winkler
CO	30.6	63.6	40.3
H_2	52.8	35.0	55.0
CH_4	15.2	-	3.9
N_2 + Ar	1.4	1.4	0.8
	100.0	100.0	100.0

It is said that the true measure of useful heat is the heat available, after combustion, above 300°F. If it is broadly accepted that the product of gasification is around 50:50-CO and H_2, then such a gas would fulfil this condition.

	SCF	Btu above 300°F/SCF	Useful Heat Btu
H_2	50	261	13,050
CO	50	307	15,350
	100		28,400

The first stage in making SNG is to shift the gas to give a 3:1 H_2 to CO mixture, in this way:

$$5\ OH_2 + 50\ CO + 25\ H_2O \ \_\_\ 75\ H_2 + 25\ CO + 25\ CO_2$$

To achieve this end a CO shift plant and a catalyst are necessary. The next stage is to remove the CO_2 which will require steam and power, after which the gas will yield useful heat to this degree.

	SCF	Btu above 300°F	Useful Heat Btu
H_2	75	261	19,575
CO	25	307	7,675
	100		27,250

Therefore after all the effort, money, steam and power a gas results processing 96% only of its original heat. Methanation to Synthetic Natural Gas carries the slide still further.

$$75\ H_2 + 25\ CO \ \_\_\_\_\ 25\ CH_4 + 25\ H_2O$$

This process, involving a very sensitive catalyst possessing a limited life, is carried out in a reactor. First, sulphur removal must take place, of itself a very expensive process with the object of protecting the sulphur-sensitive catalyst. This reaction is highly exothermic requiring removal of the low grade heat.

This process has not been put into commercial use, although it has worked successfully on a pilot plant scale. The heat yield after all this effort is only 75%, as can be seen from the following.

	SCF	Btu above 300°F/SCF	Useful Heat Btu
CH_4	25	861	21,525

This is the yield from the original 28,400 Btu's.

So if progression shows a heat loss what of the value of methane as a feedstock to the Chemical Industry? Basically this form is preferred for HCN only, but the use of natural gas for this purpose is discussed in chapter 10. So far as thermal efficiency goes there is little merit in producing CH_4 synthetically, based upon the facts to date.

The next stage is to examine types of gas.

Types of Gas

Clearly there are three types of gas classified according to their Btu values.

High-Btu Gas

This is essentially methane, which when made from coal is termed SNG - synthetic or substitute natural gas.

Medium Btu Gas

This gas gives more Btu's per ton of coal. It is probably the best fuel as well as the best feedstock for the Chemical Industry.

Low Btu Gas

This type of gas - producer gas - is diluted with about 50% nitrogen from the air used in the gasifier. It's main value is local, as a fuel or as is used in The States to generate electricity.

GASIFIERS

As has already been mentioned there are three gasifiers only which are proven. Some description and comparison between them may be helpful.

Lurgi Gasifier

This gasifier which is illustrated in Fig. 31, consists essentially of a pressure vessel containing a fuel bed fitted with a top coal feed-lock hopper and a bottom ash discharge lock hopper. The unit operates at a pressure of 31 atmospheres. It uses the counter flow principle: steam and oxygen are fed to the base while the coal is introduced through the top. Rotating distributers and rabble arms ensure even distribution of the coal. The grate also rotates as well as supporting the fuel bed. The fuel bed is surrounded by a water jacket within the main pressure shell where the steam is raised then being added to the gasification steam feed. The gas emerges from the unit at 450-600°C and is then scrubbed, when the temperature falls to 200°C before passing to the purification and treatment stages. This stage is then followed by methanation.

Lurgi has a high thermal efficiency with a gas yield equivalent to 90% of the potential heat in the coal. Even so, it has a number of disadvantages:

- size of the unit is limited and therefore output is low.

- characteristics of coal feed are limited.

- considerable excess of steam is required.

- methane produced by hydrogenation is limited by the low partial pressure of hydrogen resulting from the high volume of steam and carbon dioxide.

- oxygen consumption is high.

There is considerable scope for modification.

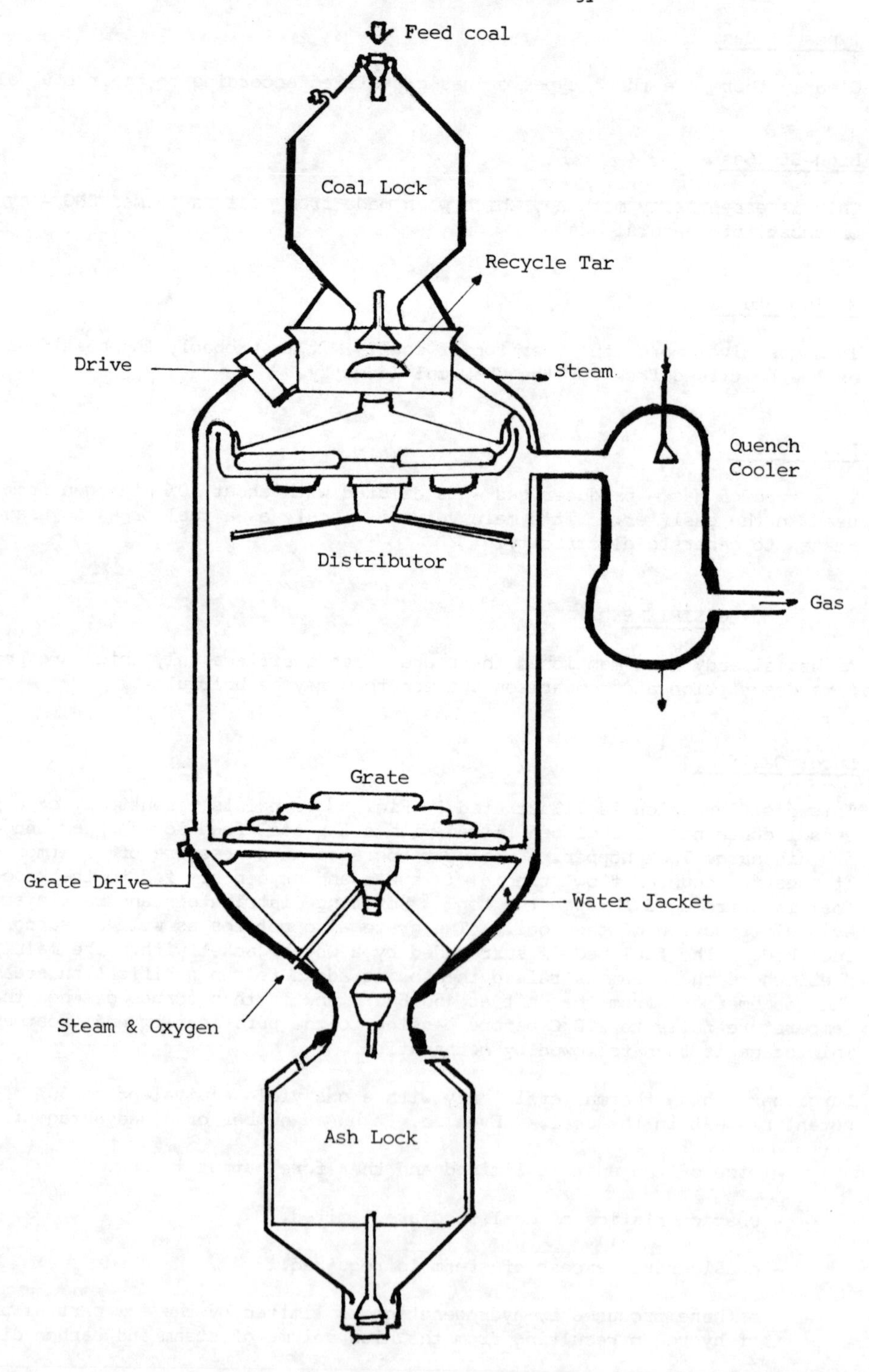

Fig. 31 The Lurgi Gasifier

Successful experiments have been carried out in Britain by employing the system under slagging conditions. Instead of the grate, a refractory hearth was introduced with steam and oxygen fed in through the surround to the hearth with no excess steam. Temperatures of around 2000°C were achieved, the ash melting into liquid slag.

The main advantages are claimed to be:

- increased output resulting in less units being needed.
- application of a wider range of coals due to elimination of ash fusion.
- greater efficiency.

There were also certain disadvantages namely:

- a slightly lower methane yield.
- increased oxygen consumption.

Traditionally Lurgi has been applied to lignites and low caking coals. However the trials at Westfield, Scotland have shown that a whole range of coals including bituminous caking coals brought over from the States can be used.

Lurgi is considered to be the best method of producing Synthetic Natural Gas (SNG).

Koppers-Totzec

This method is based upon partial oxidation of pulverised coal in suspension, with oxygen and steam. Only gaseous products are said to be evolved with no tars, phenols or condensable hydrocarbons being formed.

Winkler Process

To date thirty eight Winkler gasifiers have been built to provide fuel gas, hydrogen and synthesis gas. A further twenty six replicas are said to have been built in the U.S.S.R.

Characteristics are:

- lignite is generally used as a feedstock but coke, semi-coke, bituminous coal and caking bituminous coal have also been used.
- capable of using most forms of coal except highly caking coals. Ideally suited to relatively young geological age groups such as lignite and semi-bituminous coal.
- coals with a Free Swelling Index up to $2\frac{1}{2}$ are suitable. Above this figure they tend to exhibit lower reactivity.
- anthracites and semi-anthracites also exhibit a low reactivity and therefore do not make good feedstocks.
- coal need not be dried.
- run of mine coal regardless of fines but crushed so as not to exceed $\frac{3}{8}$" is suitable without pulverising or briquetting or specially screened flows

to individual gasifiers.

- high or variable ash contents do not affect the gasifier.
- great flexibility between maximum and minimum flows including an ability to bank up when no gas is required.
- a simple empty reactor keeps maintenance costs low.
- operates on air or oxygen.
- high residence time for solids guards against oxygen slip.
- tar does not run down the walls forming an explosive mixture with incoming oxygen, as in a fixed bed gasifier.
- tar and phenol are not formed; the effluent is low in BOD.

Clearly medium Btu gas has much to commend it, but it is important to examine the national development programmes to gain an overall view of progress.

NATIONAL PROGRAMMES

A number of countries have fairly extensive programmes in hand but the largest of these for which information is published is that of The States.

UNITED STATES

A guide as to government involvement in The States can be judged from the 1978 funding through the budget allocation as requested by the then ERDA - now the Department of Energy. Basically, the only commercial venture seen to be likely before 1985 involves Low Btu gasification, the programme for which is shown in Table 22 where the sum allocated is $65,500.

TABLE 22 U.S. Low Btu Gasification Programme

Project	Product	($000s) 1978 Funding
1. Fixed Bed Merc Morgan Town, Washington, Va. Grand Forks - North Dakota.	Stirred. Slagging.	4,500
2. Fluidised-bed	Two stage. Three stage.	5,000
3. Powerton combined cycle test facility. Pekin, Ill.	Gasifier-Lurgi with gas turbine modifier for Low-Btu gas.	28,000
4. Hydrogen from coal facility.	T.B.D.	18,000
5. Gasifiers in Industry (Six awards)	Industrial Commercial Institutional	10,000
		$65,500

It can be seen clearly from Table 22 that the combined cycle test facility is the main beneficiary.

The main objectives of this programme are:

Fixed Bed Gasification

This process is carried out at atmospheric pressure.

a) improved coal feeding, ash removal, stirring and lock hoppers. A Scrubber and H_2S removal system will be provided to clean the gas for use in combustion and turbine blade corrosion tests. A water purification system is also to be installed.

b) the development of a slagging fixed bed gasifier to examine the suitability of western-low sulphur coals.

Fluidised Bed-Pressurised-Gasification

Here three projects were set up - run by Bituminous Coal Research Inc. of Monroville, Pa.

a) a three stage gasifier - a demonstration project.

b) a two stage pressurised fluidised-bed gasifier capable of using caking coals without pretreatment.

c) a fast fluidised-bed operation sponsored by Hydrocarbon Research Inc., Trenton New Jersey.

Entrained-Bed Gasification

This is a pilot plant project at Windsor, Connecticut, to be mounted in two phases.

1) feasibility studies - already completed.

2) a pilot plant construction to provide a gasifier capable of operation at a heat rate approaching a conventional coal fired generating plant - 10,200 Btu/Khr.

Combined Cycle Test Facility

This will be the Powerton project - Pekin, Illinois. It involves gasification processes using low Btu gas in combined cycle power plants followed by gas purification units and gas turbine electrical generating sets. 500 to 800 tons of coal consumption per day is visualised, the gas being used to drive a combustion turbine for power generation. Second generation high temperature combustion turbines are to be added.

Molten Salt Pressurised - Demonstration Unit

This project will be mounted by Atomics International at Santa Suzanna - California. Coal is partially burned and then gasified completely in a bath of molten salt.

This enables high sulphur caking coal to be gasified with the sulphur and ash retained in the melt. The salt particles remaining in the gas are removed before the gas is used.

Hydrogen from Coal

This project attempts to obtain sources of hydrogen from coal, in other phases of gasification and liquefaction for industries at present using Natural Gas as the source. A daily rate of 200 tons of coal will produce 30 tons of hydrogen. The total cost is of the order of $87 million.

Ash Agglomeration

Char and fine coke particles involving gasification under ash agglomeration conditions will be used for the main part of this project, which is an extension of the previous work on the atmospheric fluidised bed.

Hot Gas Clean Up

The clean up with molten salts as a medium for gas scrubbing requires the removal of both particulates and sulphur. The need is for a continuous method of removing the impurities from the molten salt.

This can also be achieved through the use of solid absorbents for sulphur compounds which possess a greater removal capacity and a longer operational life. Also greater strength and stability in the regeneration cycle, particularly at elevated temperatures. The introduction of a fluidised bed is being examined. Iron Oxide is used as the main absorbent here.

Gasifiers in Industry

Six gasifiers are proposed with the object of:

a) demonstrating low Btu gasification technology by integration and evaluation of coal gasification systems under a range of operating conditions.

b) analyse technical data on component and system operating parameters to assist in design and development of advanced prototype coal gasification systems.

It is of interest to note the position of medium Btu gas in connection with combined cycle project. It is well established that using current technology, a gas turbine power generator followed by a steam boiler and steam turbine power generator - commonly called a combined cycle - can be more efficient than a conventional condensing steam turbine.

The main problem is that the gasifier must produce a sulphur free and particulate-free gas at an optimum pressure using air as an oxidising agent.

The Koppers process can only use oxygen. The Lurgi gasifier is currently the most advanced process. It has a considerable track record operating at the required pressure, with a development plant at Lunen. The scale at Powerton will be even larger.

There is the problem - common to fixed-bed gasifiers - of tar, phenol and ammonia as well as other effluents being produced. In addition, Lurgi is reported to produce a gas flow which is continually changing, as to rate and composition, due to the batchwise operation of the coal locks.

Westinghouse has operated a small pilot plant using a fluid bed said to require many more years to develop.

Winkler appears to be the only available established process able to meet the criteria considered necessary for combined cycle coal gasification.

The optimum pressure for a Winkler combined cycle gasifier is in the range 200 to 250 psi, higher than guarantees demanded. The conversion potential of coal into power at 33 to 45% efficiency as compared to a range of 28 to 33% should attract financial backing for a plant already designed to achieve the higher pressures required.

High Btu Gas

The allocation of funds for 1978 were made as shown in Table 23. Commercial application of the developed technology is aiming at mid-term 1985-2000.

TABLE 23 U.S. High Btu Gasification Programme 1978

	Project	Product	(\$000's) 1978 Funding
1.	Hygas Chicago, Ill.	High Pressure multi state Hydro-gasification	Nil
2.	Bi-gas Homer City, Pa.	High Pressure entrained-flow multi stage	11,000
3.	Synthane Bruceton, Pa.	Mid-Pressure fluidised-bed single stage	14,500
4.	Steam-Iron Chicago, Ill.	Hydrogen generator	5,700
5.	Hydrane	Single stage non catalytic hydro gasification	3,500
6.	Catalytic gasification pilot plant	Catalytic gasification	3,500
	Grand Total		\$38,200

Status

1. - 4. Operational

5. & 6. Feasibility Design

From Table 23 it can be seen that the major funding of the Hy-Gas project is passed. Now, Synthane and Bi-Gas enjoy the lion's share of the allocation. A number of these processes will now be described and discussed so that their relative merits may be assessed.

Hygas

This project was part of the ERDA/AGA programme carried out by the Institute of Gas Technology. A pilot plant with a daily capacity for converting about 80 tons of coal to 1.5 million cubic feet of gas has been in operation in Chicago, Illinois for some time. Since mid-1974 when the steam oxygen gasification stage was integrated with the Hygas Process to incorporate the first char-based hydrogen generation method to be combined with Hygas, two significant advances have been made:

a) pipeline quality gas was produced from Montana lignite using only steam and oxygen.

b) use of pretreated Illinois No. 6 bituminous coal.

One of the auxiliary programmes in the Hygas project was the operation of an ash agglomerating gasifier to selectively remove the ash from a non slagging fluidised bed at temperatures around 1900°F.

The present objective with the pilot plant is to obtain the maximum methane yield through high pressure operation within the gasifier.

Bi-gas

ERDA/AGA were also behind this programme, a process developed by the Bituminous Coal Research Inc. This is a pilot plant, the only slagging being developed with the object of making maximum use of the carbon in the coal and at the same time reducing the volume of waste products produced. This method includes an entrained bed involving rapidly moving char particles in an entrained state, used to dilute the coal particles and prevent their agglomeration. The project is mounted in Homer City, Philadelphia.

Synthane

The Pittsburgh Energy Research Centre developed this process, funded entirely by ERDA, at Bruceton, Philadelphia. The present project is pilot scale. It is capable of using any type of coal including highly coking bituminous types. Since 60% of the methane is formed in the gasifier no further methanation is required. Less oxygen is also needed. Figure 32 illustrates the flow sheet for the Synthane prototype process.

Steam Iron

This project offers a potential supply of hydrogen for Hygas and is carried out by the Institute of Gas Technology. This process uses char residue from any gasification scheme.

A continuously circulating stream of iron ore is cyclically oxidised and reduced in the two parts of the steam iron system. The reducing gas for converting iron oxide to iron is generated by converting residual char from the coal gasifier with air and steam into a high pressure producer gas. The iron from the reducer is then oxidised with steam to provide a stream rich in hydrogen which is later returned to complete the cycle. The system is operable over a wide range of pressures.

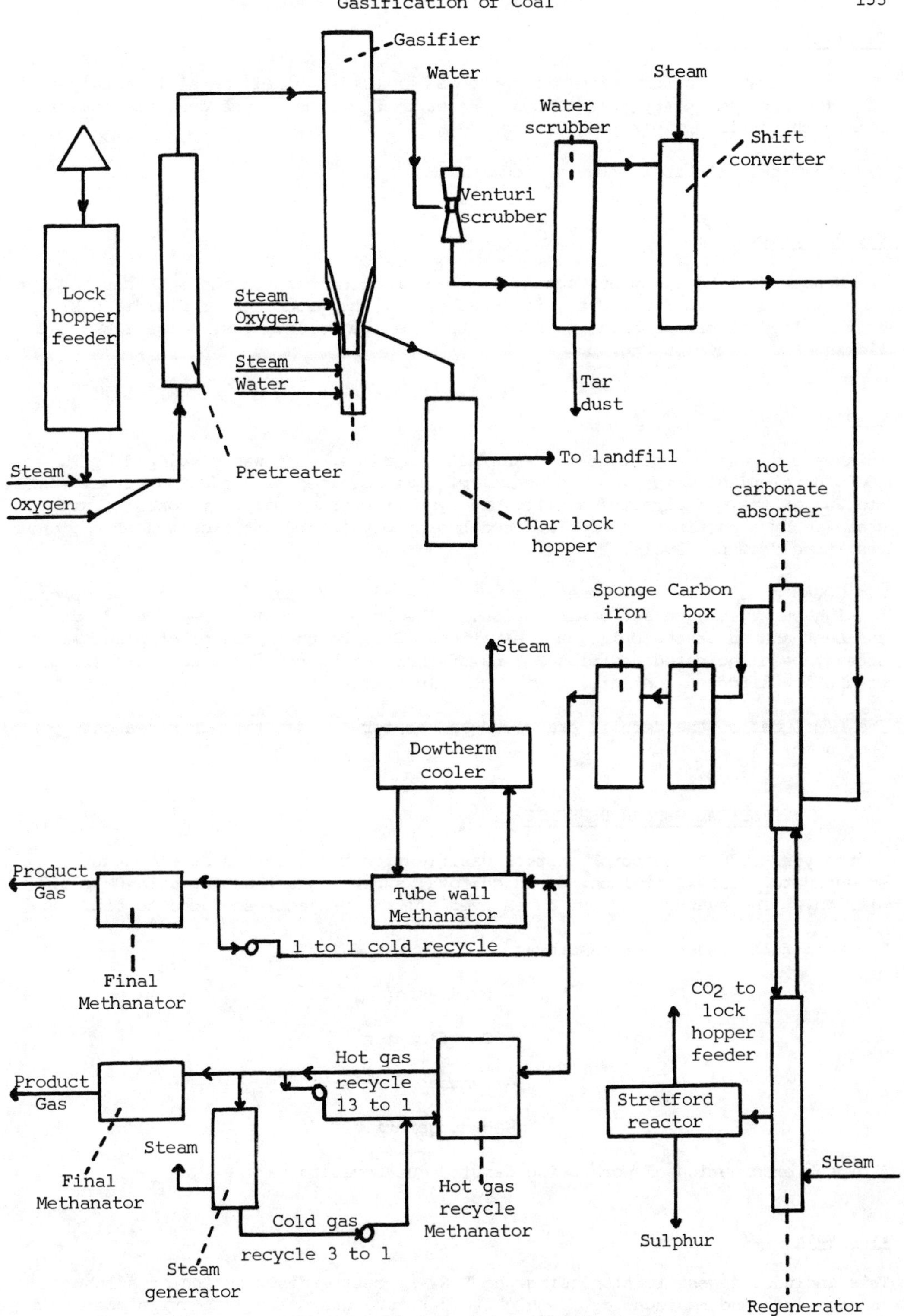

Fig. 32 Flowsheet of Prototype Synthane Process

Catalytic Gasification

This new advanced alternative pilot process involving alkali metallic catalysts is being carried out by Fossil Energy Research Division. The coal is converted within the gasifier in a single stage.

Two other projects are worthy of comment.

CO_2 Acceptor

The Conoco Coal Development Company has been responsible for the work carried out at a pilot plant in Rapid City, South Dakota. The process is a fluidised bed to convert lignite or sub-bituminous coal to pipeline quality gas. The schematic flowsheet is shown in Fig. 33.

Cogas

This process was not part of the ERDA/AGA programme. It was developed by the COGAS Development Company - a consortium. It is aimed at producing either medium Btu Gas or that of pipeline quality and synthetic crude oil from coal. This is achieved by a combination of fluidised bed pyrolysis and gasification of the char resulting from pyrolysis.

The COGAS process is a low-pressure process - 45 to 75 psi - using air as opposed to high-purity oxygen for gasification. The original tests were carried out at the Leatherhead Laboratories of the National Coal Board. The pilot plant as has already been indicated gasifies 2 tons of char per hour to produce synthesis gas with a low nitrogen content and without using oxygen.

The remainder of the details are given in chapter 4 under the NCB research programme.

FEDERAL GERMAN REPUBLIC

In the year 1976/7 the total cost of gasification research and development amounted to DM 264.2 million, of which total support grant amounted to DM 133.8 million. The average support taken over all the projects amounted to 51%.

The work falls under four headings:

- Pyrolysis
- Gasification
- Synthesis
- Hydrogenation

A closer examination of work being carried out reveals:

Pyrolysis

This includes degasification using the L.R. process - distribution of sulphur in intermediate and residual coke - high temperature gas-solid suspension cooling in heat exchangers.

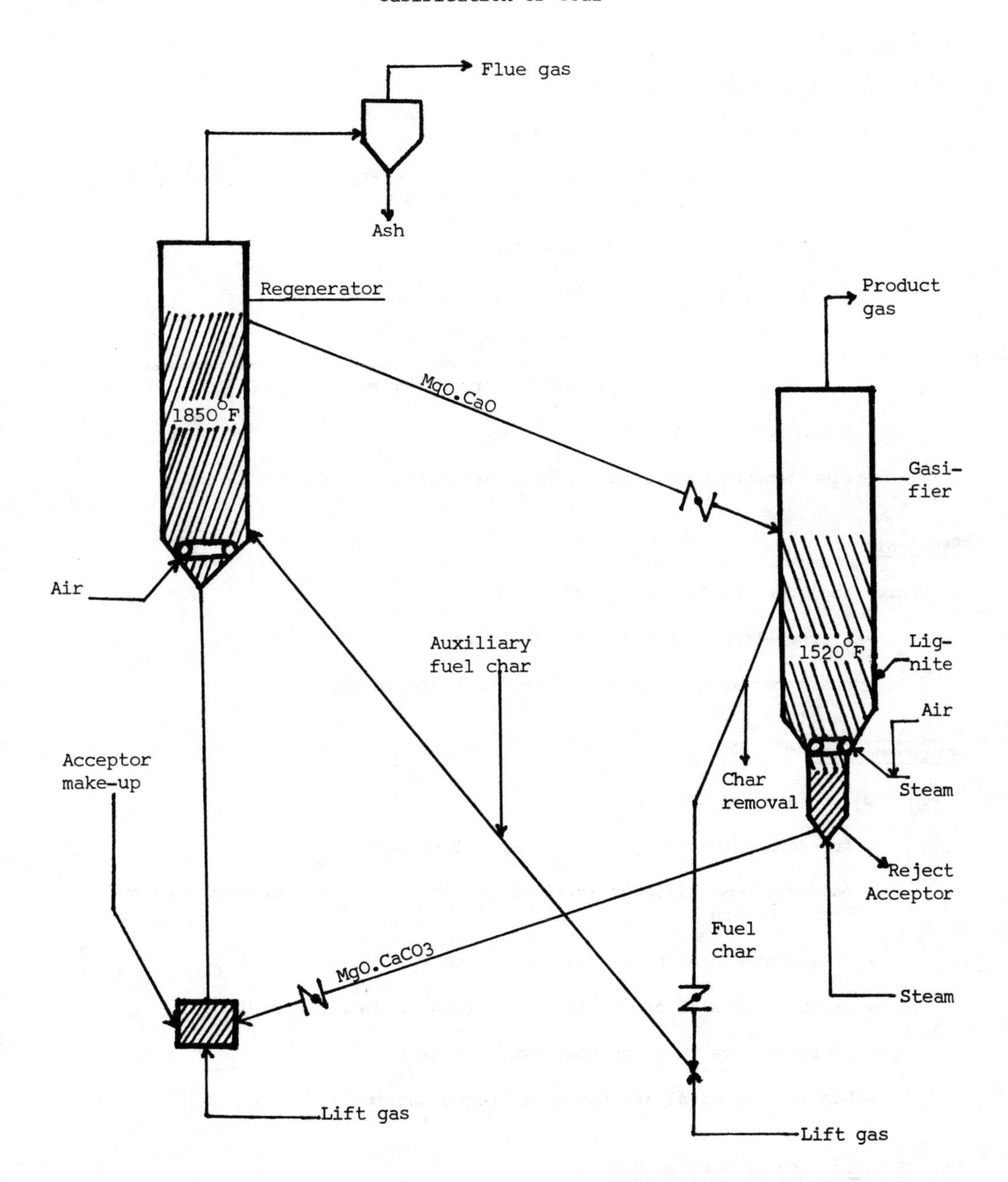

Fig. 33 CO_2 Acceptor process diagram

Gasification

There are fifteen projects including:

- pressure gasification of pulverised coal.
- production of synthesis gas by pressure gasification of coal dust with oxygen and water - by the Texaco Gasification Process.
- Lurgi process for synthesis/SNG.
- gasification in a) high temperature Winkler gasifier
  b) tube furness gasifier
  c) from nuclear process heat
- methanisation in a fluidised bed.
- simultaneous conversions and methanisation of CO-rich gases.

Synthesis

This area includes the following projects:

- Fischer-Tropsch process - catalysts
- basic research on catalysts for CO hydrogenation

Hydrogenation

Seven projects are included in this field:

- broadening the raw material base - two projects
- physico-chemical investigations of bituminous coal extracts and their fractionation
- treatment of tar from coal conversion
- demonstration plant for the production of coal oil in U.S.A.
- improved catalysts for coal hydrogenation
- homogeneous catalysts for coal hydrogenation

Coal Gasification by Nuclear Heat

In conventional gasification processes only 60-70% of the coal is converted to gas, the remainder having to be burnt in order to provide the necessary energy. One of the above projects involves the use of process heat from high temperature nuclear reactors (HTR) as a source of energy, so transforming the coal completely to gas.

This process offers considerable advantages as compared with existing processes, in particular saving coal since more gas can be produced from less coal. There are less emissions of particulates, SO_2 and NO_x as well as other pollutants, since HTR produces steam and electricity instead of coal-fired boilers.

The schedule calls for a plant capable of using process heat from the HTR at 950°C, the temperature of operation for 1985, an indication that the process should be available by 1990.

Some of the advantages claimed for SNG produced by steam gasification are:

- coal requirements per G Cal. SNG produced drops from 1.4 G Cal. to 0.9 G Cal. This is a striking argument in countries where coal is in short supply and particularly where coal must be imported.

- gas costs which rise with the cost of coal by a factor of 1.4 in the case of conventional gasification, but only 0.9 when nuclear heat is used.

- the emission of CO_2 falls from 305 cubic metres/G Cal. SNG for conventional gasification to 177 cubic metres/G Cal. SNG for nuclear when both the production and combination of the SNG are taken into account.

Bergbau-Forschung GmbH, Essen, Rheinische Braunkohlenwerke A.G., Köln and Kernforschungsanlage Jülich GmbH and Jülich are co-operating in this project.

The trial runs which took place during 1976 marked a change in this field from the study and planning phase to one of development supported by experiment. Despite many operational difficulties it has been possible to demonstrate the performance data on which the planning was originally based.

Coal gasification continues to enjoy priority in the programme with regard to financial backing supported by the many potential applications for the gas - SNG synthesis gas, town gas and low-Btu gas. Even so at present the chances are remote that economic use of these processes will be made in Germany. As has been seen from the programme, general engineering effort on coal dust gasification projects under pressure have continued, system components having been contracted out. The basic engineering for the project on the advanced development of Lurgi fixed bed gasification has been completed. The start up of this facility should now take place in 1979.

Electric Power

New developments and improvements in processes with regard to conversion of coal to electric power should make their impact, in economic terms, earlier than otherwise might have been the case; they improve the competitive position of coal. It is intended that considerable time should be devoted to this field of development. The work is aimed at conversion processes having greater efficiency and lower investment cost in respect of sulphur removal from the solid fuel. This is an integral component of the process. Three different processes will be developed:

a) The coal pressure gasification (KDV) technology where coal is gasified with air and the resulting gas is converted in a gas-steam turbine process, after purification, is underway on an industrial scale experimental facility at Lünen. It has confirmed the original results and assumptions but at the same time revealing new problem areas.

b) The VEW process (Vereinigte Elektrizitätswerke Westfalen A.G.) based upon rapid degasification of coal and the subsequent removal of sufficient sulphur and volatile matter by recognised processes. Trial runs in an experimental facility confirmed this concept which is based mainly upon accepted power station technology. Conversions to large scale industrial use appears to be feasible at low cost.

c) Fluidised-bed combustion under pressure offers the considerable advantage of sulphur removal through the addition of limestone in the combustion chamber. Involvement in the IEA project on "Pressurised fluidised combustion" will provide an opportunity of studying the basic phenomena of fluidised-bed combustion on an industrial scale. It will also assist the technical design of the combustion chamber in a commercial system. It is expected to take ten years to perfect the process, demanding conventional methods for flue clean up to ensure that the environmental regulations can be met. This need is being met through a project study to set up a demonstration plant for flue gas sulphur removal by the Saarberg-Hölter process.

Coal hydrogenation has been used industrially in Germany for over thirty years. Accumulated experience has enabled a process to be tested continuously in the laboratory under the title "New IG Farben Process". Reduced investment costs and good yield of high quality have confirmed original expectations. Plant design has also been confirmed.

Key processes for future energy supply are already being covered by parallel development efforts - gasification and conversion of coal to electric power.

The view in the Federal Republic is that coal conversion technology cannot be considered in a national context, but rather through international collaboration. Joint projects should be set up where feasible. Already five such arrangements across the board have been reached within the scope of IEA.

BRITAIN

Mention has already been made in chapter 9 of the British coal research effort. There are two main programmes, the Westfield Gasification Project and the Grimethorpe Fluidised Bed.

Westfield Project. This provides an opportunity for British Gas - an entity quite separate from NCB to exploit its technology and expertise in the manufacture of gas from coal and oil. This is happening by licencing the CRG process for the production of SNG from naptha and by making available its coal gasification technology plant etc. for further development under U.S. sponsorship. Some fourteen streams of CRG based SNG plants were installed in the U.S.A. producing 33.34 x 10^6 cubic metres per day of SNG.

The Westfield Lurgi gasification plant has now been changed to a gasification development centre, which has a number of projects behind it.

The object of the exercise has been to obtain experience and data which would enable process guarantees to be given for a range of coals previously held to be unacceptable.

The operation covered a range of coals: it might be helpful to examine some of the properties of some of these coals.

Western sub-bituminous coals. These are characterised by their relatively high

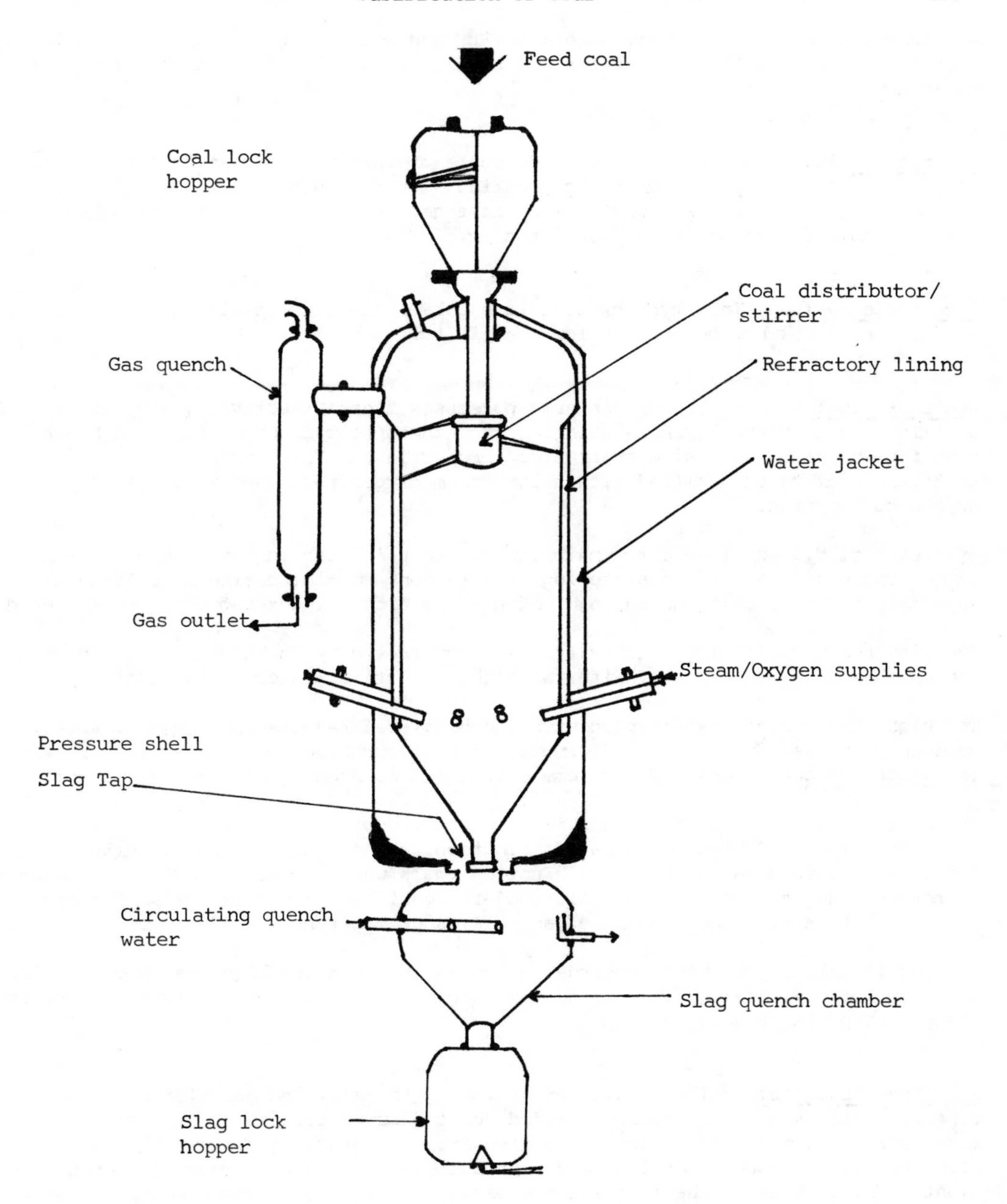

Fig. 34 The Westfield Slagging Gasifier

moisture content, low sulphur, high activity and non-caking properties. In theory they are an excellent feedstock for the Lurgi gasifier, but for a rather low fusion point ash.

Illinois coals. These regions provide coals of fairly low moisture content, but they are high in sulphur and strongly caking - or reasonably so. But their ash fusion temperatures are somewhat low which together with low reactivity makes them less suitable for normal Lurgi gasification.

Pittsburgh coals. This type of coal is highly fusible and swelling. They are also strongly caking, but have a low reactivity.

Westfield coals. Coal from this area possesses moderate activity, very low swelling and is virtually non-caking. The ash content is high, but also has a high fusion point which gives refractory properties. These properties enable Westfield coal to be gasified with a low steam oxygen ratio and a relatively low oxygen consumption.

By contrast, U.S. coals due to their low fusion point ash require a greater steam oxygen ratio to prevent ash sintering, limitations in time making it difficult to adjust operating conditions which might have permitted lower ratios to be achieved.

The Pittsburgh coal's low reactivity was demonstrated by the high oxygen consumption which contrasted markedly with the highly reactive Western sub-bituminous coals.

The high oxygen ratio experienced with the U.S. coal results in high yields of gaseous liquor equivalent with Illinois and Pittsburgh coals to in excess of twice the weight of coal used. It is more than twice as much as with Westfield coal.

The most useful fact to emerge was the suitability of the Western sub-bituminous coals for fixed bed high pressure gasification, as well as performance data. The trials also gave data on the use of Eastern coals which enabled Lurgi to calculate performance guarantees and the ability of the gasifier to handle coals of a size similar to that of run of the mine, as well as caking types.

The results all represented important advances in fixed bed high pressure gasification technology, greatly increasing the range of coals which this type of gasification is capable of using.

Slagging conditions. The production of SNG by the slagging gasifier route need differ little from the conditions needed for the Lurgi gasifier. Tar and oil must be removed from the crude gas emitted from the gasifier followed by CO conversion, removal of carbon dioxide, hydrogen sulphide, organic sulphur and a range of light hydro-carbons in the Rectisol processes by a low temperature methanol wash. After this the purified gas is passed through a methane synthesis unit. Here most of the oxides of carbon are hydrogenated to methane in the presence of a catalyst.

A diagrammatic process-route is shown in Fig. 34.

Oxygen consumption has attracted particular interest. A higher oxygen consumption was expected on the slagging gasifier as compared with the Lurgi results during the U.S. Coal trials. This can be seen from Table 24 to have been generally confirmed except in respect of Western coals.

TABLE 24 Comparison of Oxygen Consumption Figures using a variety of Coals under Slagging and Non-Slagging Conditions

Conditions	Coal	Steam-Oxygen Ratio mol/mol	Oxygen Consumption m^3(St) per 100 MJ crude gas
Non-slagging	Western Types	9.1	1.13
	Illinois 5	8.6	1.67
	Illinois 6	9.8	1.69
	Pittsburgh	9.85	1.85
	Westfield	6.2	1.71
Slagging	Donisthorpe	1.10	1.45
	Newstead	1.06	1.49
	Frances	1.15	1.43
	N. Dakota	1.11	1.43

A demand for coal-based SNG in Britain is unlikely to occur before the 1990's, although peak shaving SNG may be needed earlier.

It will be of interest to note the Sponsoring Companies, all of which are American, backing the Westfield Slagging Gasifier Project.

Continental Oil

El Paso Natural Gas

Gulf Energy & Minerals

Michigan Wisconsin Pipeline

Natural Gas Pipeline

Panhandle Eastern Pipeline

Cities Service Gas Company/
Northern Natural Gas

Standard Oil of Indiana

Southern Natural Gas

Sun Oil

Texas Eastern Transmission

Tennessee Gas Pipeline

Transcontinental Gas Pipeline

Electric Power Research Institute

Otherwise, British activity is based upon a three year project, following a proposal to the European Coal & Steel Community in 1974 which agreed to 60% funding.

The main need is for turbines to be developed which will operate at 1250°C as compared with the present 1000°C, so that more efficient power generation may be achieved.

The main requirements are that the bed should:

- be able to use normal power station coals.

- run at pressures corresponding to the turbine at 20 atmospheres.

- be a cheap process

The three proven methods of gasification already discussed in detail have been examined, with a trend towards the entrained bed type. Very fine coals can be used with oxygen blown in, although there is a tendancy to temperatures which are too high. The ash emerges as a molten slag which makes heat recovery difficult. Sodium and Potassium in the gas must be removed, since their presence corrodes the turbine blades.

A new concept is being developed which is pressurised to include combined cycles introducing a gas turbine expected to be 30% efficient and a steam turbine of equal efficiency (on the remaining 70%).

Europe

Little work is being carried out in the rest of Europe on gasification. None was sponsored by the Commission in 1976, while Belgium's main interest is "In Situ". The Commission have made available 20 m.u.a.s annually for gasification.

France

Work involving HT reactors totalling .02 million units of account was carried out in 1976.

Italy

A feasibility study took place.

Netherlands

Assessments costing .05 million units of account were made during 1976.

Without doubt the United States is the leader in this field but as one of the main witnesses informed the House of Representatives Committee on Appropriations; "One of the things we are learning is how not to do it". In this difficult field that could be progress. Others may save their cash by looking on for the time being.

IN SITU GASIFICATION

It is a surprising fact that despite numerous pilot scale tests, which include the British attempts by Humphrey & Glasgow in the early fifties, no commercial scale installation has worked anywhere in the world apart from the Soviet Union, but even there the enthusiasm has declined in recent years.

The reason for ceasing this effort was mainly economic accompanied by an abundance of energy which was cheap. Open cast coal in particular made the need less urgent.

However, following the 1973 energy crisis a number of countries, Belgium, Canada, Germany and the United States, have turned to means of recovering otherwise unmineable coal - Underground coal gasification (UCG).

The really important aspect is to predict the cost of any gas produced against prevailing economic conditions.

U.G.C.

The basic concept is not complicated. Gases can be produced from coal when it is heated, the composition depending upon the manner of heating. In the absence of air the coal is carbonised to produce coal gas, tar, coke and ash. The resulting gas has a calorific value of about 540 Btu per cubic foot. This is not repeated in UCG, the tarry by-products condensing out in the coal seam retard the reaction. The irrecoverable coke residue is left behind reducing the general efficiency of the system.

But if the coal is heated directly by its own combustion in air the residue consists of ash only. This involves a number of complex reactions. Carbonisation, distillation, cracking and combustion of the by-products are all involved in practice. However, the process may be represented in this way.

$$C + O_2 \longrightarrow CO_2 + \text{heat}$$

$$2H_2 + O_2 \longrightarrow 2H_2O + \text{heat}$$

If the hot gases remain in contact with the coal so far untouched, once all the oxygen has been used up, a second set of reactions occurs.

$$CO_2 + \text{heat} + C \longrightarrow 2CO$$

$$H_2O + \text{heat} + C \longrightarrow CO + H_2$$

These last reactions are endothermic, making use of the sensible heat from the gas to produce a gas capable of combustion which can be used as a fuel. The calorific value of this gas will depend upon the relative proportions of CO and H_2, as we saw earlier, when discussing the merits of medium Btu gas. It should be around 200 Btu per cubic foot. But if air has been used instead of oxygen then the nitrogen contained in the air dilutes the product gas and at the same time reduces the calorific value to below 100 Btu per cubic foot.

Sufficient background experience has been collected to see that there could be a future for burning coal underground to obtain gas. It has a considerable potential for producing large quantities of gas for electricity generating stations or for use as a low Btu gas in industry. Some countries see it as a means of producing SNG although as has been shown earlier this end product is both expensive and wasteful in energy use.

Those technologies employing underground mining operations on a large scale, over the post war years, have generally been abandoned. The object is to recover the energy stored underground, while eliminating the need for men to face the problems and hazards of working below ground.

The technology used in various countries where work is in progress or experiments have been carried out is based upon what is called the filtration method.

Two holes 30 to 80 metres apart are connected by a channel made in the coal seam by making use of the natural permeability of the deposit. Alternatively, water or air may be injected at high pressure. This is called a linking operation. The fissure produced is made larger by injecting air to provide a combustion area running counter current through the coal. This results in much greater quantities of air being injected so that forward gasification can proceed as the gasification front moves in the direction of the gas flow.

Constituent Process of Gasification

Basically, the deposit is made up of three component factors.

Oxidation zone. Here, the coal reacts with the oxygen of the gasification medium to produce CO and CO_2, a substantial amount of heat being released in the process. This heat is transmitted to successive zones by the gas current, the coal and surrounding rock.

Reduction zone. This process commences as soon as the oxygen of the gasifying media has been used up. The Carbon Dioxide and water vapour entrained by the gas flow are almost completely reduced to CO and H_2.

Distillation zone. Here the coal is heated by convection along the path of the gas flow and by conduction in the area surrounding the high temperature zone. Volatiles become released which then mix with the gas and in the process, enriching it.

The problem encountered underground is that these three stages occur in a random fashion. Gas flows cannot be controlled, due to changes in the resistence of circuits and the difference in effective surface areas available for the gas/solid reactions. Gasification reactions could be slowed up by temperatures which are too low.

Energy Potential of the Deposit

Again there are three sources:

Firedamp. This accounts for some 3% only of the total calorific value, which would be present at about 25 cubic metres per ton. On the other hand that from adjacent seams may be released by the fracturing of the strata. As has been seen earlier, degasing of the adjacent strata is immensely important, particularly that of the deep lying coal.

A considerable amount of data has been collected on the quantities of firedamp which can be made available. At Klorenthal, Germany, as much as 92.87 cubic metres/ton has been recovered through a bore hole drilled from the surface.

At the Charleroi coal field the specific firedamp recorded during methanometric measurements reached 189.6 cubic metres/ton. Equally as was shown in an experiment, again at Charleroi, during Methane drainage from abandoned shafts, very considerable quantities of gas may be released after the coal has been worked. Fundamentally, firedamp drainage presents the same problems whether worked by underground gasification or conventional methods.

Volatiles. Progressive heating of the coal, in a reducing atmosphere between 450 and 1000°C, will remove the volatiles.

Fixed Carbon. After the volatiles have been drawn off, fixed carbon remains. Its calorific value is in the region of 8000 K cals/Kilogram ignoring inerts present.

Pioneers

Britain, U.S.A. and U.S.S.R. have been the main pioneers but this is not to demote others who have been active in the field.

There is a record of Sir William Siemons discussing the general idea in the mid 1800's, while Mendeleyev followed suit in Russia some years later. A Britisher was the first to carry out experiments just before the first World War. But little happened until the mid 30's when the Russians revived it.

UCG has been seen to solve a number of problems according to the view point of the commentator.

Not surprisingly the U.S.S.R. saw it as a means of shortening the working day and the working week, as well as reducing the cost of electricity generation to a fraction of the present level.

In the Federal Republic today where miners earn about £8000 a year the object is to reduce the labour content of mining. India, of course, would not have a similar interest - for obvious reasons.

Britain

Originally UCG was the responsibility of the Ministry of Fuel and Power transferred in 1956 to the National Coal Board. The first objective was to make underground gasification work. Air was pumped into the coal seam via a compressor, the gas emerging from a second vertical bore hole along the seam. The first trial began from an open cast working. A hole was drilled to pass through the seam and in the plane of the seam, passing across the bottom of a number of vertical holes one of which was used as an inlet for the air with a further hole acting as a gas escape.

Later in the absence of an open cast site from which to break,in access roads in a mine formed the boundaries of the seam. Vertical holes were bored at either end to introduce the air and allow the gas to escape. High pressure air was supplied to one vertical bore hole which percolated through the seam. The passage through fairly hard coal was very slow.

Electro-linkage was also tried in which two electrodes were fitted into the seams at the bottom of vertical bore holes and a high voltage applied across them. It was not successful.

This was followed by the blind borehole method. It consisted of a central drilling gallery from which holes were drilled - 250 feet into the seam. The end was blind. A steel pipe was inserted into the hole to conduct the air to the reaction zone and act as a heat exchanger. Problems arose with the steel and although successful, other methods showed greater promise. The simple idea of a single channel in which the reaction took place was not sufficient.

One more attempt was made based upon a survey of Belgian, Polish and Russian work being carried out. It involved the setting up of a primary reaction on a broad front. Gases produced were ducted through in-seam bore holes where CO was produced before emerging through vertical bore holes. Ignition was brought about by wooden sleepers, a broad reaction front resulting. A low calorie gas was produced, mainly due to the lack of thickness of the seam.

It was at this point that it was decided that UCG work would stop - on economic grounds. However, a further attempt was made between 1962 and 1966. The former working site was exposed by open casting to enable the gasification zones to be examined. High temperatures had plainly been reached in the reaction zones but for a very limited distance only - not more than 6 metres above the seam. The gasified zones in the seams were surrounded by a zone of carbonised coal varying in width from between 0.5 to 1.2 metres. The heat was localised even after the time which had elapsed since use; it was warm to the touch in parts of the zones. The carbonised coal showed progressive loss of volatiles; that adjacent to the burned face containing between 2.5 to 20% volatile matter. The cavity roofs had partially collapsed although there was evidence that the air passages had remained open.

Belgium

The Belgian attempt began in 1948 using high quality coal - itself remarkable. It involved a steeply-inclined seam with the flame front moving sideways from an inclined drift as opposed to moving upwards from the more usual horizontal plane. Some combustible gas was produced but generally the results were poor.

In 1973 when major changes of world wide significance occurred Belgium was quick to consider the recovery of energy resources contained in deep-lying seams.

INIEX was charged with making proposals; the 1948 work was re-examined together with work being carried out in other countries.

Certain factors had to be considered:

- gasification at shallow depths
- the high population density in Belgium
- fuller exploitation of the energy potential than could be achieved by methane recovery alone.

These considerations quickly led to the adoption of the method of underground gasification of deep-lying seams, at elevated variable pressures.

The principle was contained in the following summary:

Several gasification holes were drilled, to ensure access, to enter the lowest seam of the measures to be exploited. First the seams must be made permeable by the injection of air or water at high pressure, before gasification begins. Care must be taken not to fracture the roof strata by selecting an injection pressure between the hydrostatic and lithostatic. This avoids heating an unnecessarily large mass of rocks.

As soon as combustion has begun, a pattern of diagonal circulation between bore holes is set up, some to inject the air or oxygen while others enable the gas to escape. A cyclic variation of pressure in the reaction zone (between 20 minimum and 50 atmospheres maximum) is attained by varying the timing of air injection and recovery of the gas.

INIEX process. This had five main features.

1) Use of high pressure to ensure

- high gas flows in small diameter boreholes.

- greater efficiency through increased speed of gas/solid reaction and transfer.

- improvement in thermal balance through reduction in heat loss, as a result of rock heating and evaporation.

- a reduction in the amount of energy required to overcome the pressure drop in the underground circuits.

- an increase in the calorific value of the product gas by shifting of the chemical equalibria.

2) Cyclical pressure variation resulting in:

   - extension of the zone of influence of each bore hole.

   - increased fracturing of the coal.

   - permeation by the gas of the caved debris.

3) Bore hole coding and recovery of the sensible heat of the gas.

4) Combination with firedamp drainage of higher seams.

5) Electricity generation.

No review of any of the basic decisions has been necessary in the last two years.

Options on Types of Gas

The view was taken that high-pressure underground gas generation could be used to obtain four types of gas according to the combustion medium introduced.

a) air only - producer gas

b) alternating air and steam - water gas

c) oxygen and steam - synthesis gas

d) hydrogen-rich gas - high calorific value gas

However the study will concentrate on the generation of producer gas by injection - according to the Belgian Director General of Mines.

German Agreement

The Belgian and German governments both possess large underground coal reserves which cannot be recovered economically by conventional means.

Both have been involved for some years on research and development work carried out in their respective countries on in-site gasification of deep lying coal on lignite deposits.

The German programme, as was seen in chapter 4, is directed towards the production of synthesis gas or SNG for mains distribution through underground gasification. Belgian research is of course aimed at electricity generation using gas produced by air injection.

The complementary nature of the objectives has led the two governments to co-ordinate their R & D to ensure no duplication of effort. They seek to establish the feasibility of underground gasification at high, variable pressures and to investigate technical and economic conditions surrounding the use and industrial application of underground gasification process.

Exchange of information will cover:

- work programmes
- preliminary and laboratory studies
- interchange of staff
- joint full-scale experiments in-situ
- joint design and construction of pilot plant for treatment and upgrading of gas

The agreement was signed in Brussels in October 1976, making provision for accession of other states willing to co-operate provided that no delay to the main objectives resulted.

It is perhaps useful to note the large investment programme to which both countries are already committed although reference to the German effort in the general context of coal was made in chapter 4.

TABLE 25 Comparison Belgian & German In-Situ Programmes

	1975/6	1977	1978	Total	Belgian Francs
Germany DM 10^6	1.5	4.7	4.8	11	165
Belgium Bfrs. 10^6	30	40	148	218	218

United States expenditure over two years 1977 and 1978 is the equivalent of Bfs. 700 x 10^6. As was shown in Table 14 the greater part falls in 1978.

U. S. S. R.

As has already been indicated Russia has played a pioneering role in this field with three plants operating continuously for the past 20 years or so, in deposits of coal and lignite at depths varying between 100 and 300 metres.

Despite regular production indicating that some of the technical problems have been overcome the quantities of gas actually produced were small as compared with the hopes of the late 50's.

According to a Belgian working party, which made study trips to the U.S.S.R., there are three main plants said to be operating in the Soviet Union. These plants producing dissimilar quantities of gas are said to have annual outputs expressed in million x 10^3 cubic metres stp, and are sited in the following areas of which the Angren complex is the largest:

- Yuzhno-Abinsk (Kuzbass) 0.4
- Angren (near Tashkent) 0.6
- Shatsk (Mosbass) 0.2

For U.S.S.R. as a whole annual output produced by underground gasification is in the region of 1.5 million x 10^3 cubic metres stp. Assuming a calorific value of 800 to 1,000 k cal/cubic metres stp, this is equivalent to 1.35 million G Cal per annum or 200,000 tce at 7,000 k cal/kg.

However there are limitations. The gas must be consumed on-site, while its cost at around 3 roubles per G Cal cannot compete with either natural gas or open lignite mining.

A breakdown of costs at Angren, reveal:

- compressed air 33%
- labour 25%
- materials, plant and depreciation 25%
- overheads 17%

Boreholes account for a small percentage only of total costs since about 80 to 150 are drilled in a year. However, costs may be misleading when it is remembered that the minimum income in U.S.S.R. is sixteen pounds a week, while the average, including salaried people, is about forty five pounds a week. On the other hand rents have remained unchanged since before the war and goods as well as services are priced according to State policy, as opposed to market forces. The rouble being non-convertible enables inflation to be controlled almost totally. Average production per borehole is about 4 to 7.5 million cubic metres stp of gas, equivalent to an effective gasification of 1,500 to 2,800 tons of coal.

Geology & hydrology. Gasification at shallow depths is dependent upon local hydrology.

- the need to prevent accumulation of water in the underground gas generator.
- conservation against disturbance and pollution of aquifers supplying water distribution systems.

This need is met through the following procedures:

- detailed study of the geology of deposits, size of water carrying formations, ground permeability and dip of the strata, including faults.
- division of the coal seams into panels which are fairly similar in terms of depth and hydrostatic pressure.
- controlled pressures of air injection to match the hydrostatic pressures preventing both gas leakage or an inflow of water.

Experience has shown that in practice the theoretical values do not strike the equilibrium needed. The difference is about 40% through probable losses of air either through ground fissures or the distribution network.

Linking operations. Three techniques are used in U.S.S.R.:

- water injection
- injection of compressed air
- in-seam drilling in steeply inclined strata where the seams outcrop

At Angren, this is achieved by counter current combustion, compressed air being injected at pressures of between 30 and 50 atmospheres, the distance between boreholes being between 20 and 25 metres.

Gas recovery. Less holes are used than for linking in the gas recovery process. They may be spaced at intervals of up to one hundred metres. Special measures must be taken to protect these holes from possible subsidence in the gasification zone since they may be needed for a considerable period. This is best attained by placing the holes on the perimeter of the coal panels or by using deviated boreholes.

Gas quality. Gasification efficiency as well as quality of the gas depend upon a large number of factors including:

- thickness of the seam
- type of coal
- water encountered
- flow of gas

Scale is important for stability of gas composition and flow rate, which in the view of the Russian technicians should be not less than 4-5 hectares.

Gas cleaning. The gas turbines used at Shatsk for electricity generation have gone out of service. All the gas produced is burnt in steam boilers or nearby factories. Burners of simple design have been developed to deal with ultra low-calorific value gas, but at Angren generating station the gas is injected at the same time as pulverised coal.

The gas has a low dust content and few condensable products so that little cleaning is required except for drain seals at intervals along the pipes with washing and cooling by means of water sprays in counter-current scrubbers with wooden screens. Accumulations of the collected impurities in the water, which is used in closed circuit for several months and precipitation of the sulphur by oxidation in the presence of air takes place in large settlement containers.

United States

Ever since 1973, the United States has been obsessed with the need for self-sufficiency. It is unnecessary to look beyond the 1978 Coal Programme Budget Allocation to the then ERDA to confirm this fact. The estimated budget stood at $384 million. Of this sum, $11 million was allocated to the In-Situ Gasification Programme, details of which are shown for operating costs in Table 26. The Hanna project is of considerable interest world wide.

TABLE 26 U.S. In-Situ Coal Gasification Programme
Operating Costs - 1978

	Major Projects	Location	$ Millions
1.	Linked vertical wells	Hanna Wyo	4.2
2.	Packed bed	Hoc Creek Wyo	2.2
3.	Deviated wells - Longwall	Pricetown W. Va.	1.2
4.	Dipping & dry beds - advanced concepts	Sundry	3.4
			$11.0

The projects shown in Table 26, are in fact different ways of achieving the same end.

Three organisations within ERDA were involved:

- the U.S. Bureau of Mines
- the Office of Coal Research
- the Atomic Energy Commission

In addition a number of U.S. Oil Companies have the same end in view. The extent of U.S. Coal reserves discussed in chapter 2 provide a good reason for this interest, but the following attractions make "In-Situ" worthwhile pursuing.

- It is not labour intensive. Men are used above ground which is more attractive than the deep mines which are constantly short of men.
- There are few environmental disadvantages as compared with the quantities of rock and dirt which have to be stowed below ground.
- The method is adaptable to a range of conditions making it possible to exploit seams which cannot otherwise be worked because they depart too far from standard, with regard to thickness, depth or dirt content.

A measure of the renewed interest may be gauged from the funds allocated to the Hanna Wyoming experiment over the period 1973 to 1980.

1973-4 first phase - 2 years - $1.5 million

1975-6 second phase - 2 years - $4 million

1977-79 pilot project with turbine 15-30 MW $23.5 million over 3 years

Application of In-Situ Methods

It is claimed that five types of fossil fuels are suitable for in-situ combustion, distillation or gasification. They are bituminous sands - coal - oil - lignite and bituminous shales. Coal and bituminous shales are to be found in abundance in the United States.

Considerable interest continues in recovery of the heaviest hydrocarbon fractions in oil begun in the 1960's. An underground linking of boreholes, some of which, as we have already seen, are used for air injection while others are used to recover the products of combustion. Once combustion has begun, the temperature rises and the heavy oils are cracked. The petroleum cake residue is burnt in-situ forming a combustion front which drives the volatile products towards the boreholes. Temperatures attained are around 600°C - usually lower. Some 100 trials are said to have taken place over the past ten years, occasionally attaining a long term industrial operation.

Use of bituminous sands is still in the experimental stage. The main difference as compared with oil is the part played by the bituminous impregnating the deposits, their viscosity and the coke residues resulting from cracking.

Lignite and high volatile sub-bituminous coals have not made very much progress.

The average reaction temperatures attained in the course of these experiments is between 800 and 1000°C. The gaseous phase contains mainly volatile, water gas and fumes from the combustion.

The process heat comes from the partial combustion of the coke residue, while varying amounts of unburnt coke remain in the ground.

Gasification of low volatile coals is more difficult due to the increased density, demanding higher temperatures to ensure that the loss of coke is not excessively large. But bituminous shales are the most difficult of all to develop using this method. First the high density surrounding rock must be fractured using mechanical means, explosives or a nuclear underground explosion.

The problems have been highlighted and rightly so. This is recognised by the position in the energy programme, where it is visualised that commercial application of the technology is scheduled for 1985 to 2000.

However, the types of deposits to which this technology may be applied best - oil and bituminous sands - are limited. Coal and bituminous shales are quite a different matter. But certain advances have made the possibilities more attractive than formerly:

- the in-situ techniques for oil recovery
- hydrofracture of rocks
- direction of drilling techniques
- use of explosive slurries
- development of turbines using producer gas

Current American Projects

The budget allocation in Table 26, indicated four current projects.

The first phase of the new American developments in the field of UCG was completed in 1968 by Gulf Research and Development Corporation in Seam 14 of the Colonial Colliery in West Kentucky.

This seam is 2.70 metres thick and lies at a depth of 32 metres, being open cast and bituminous.

The gasification was carried out 150 metres in advance of the open cut, using an air injection borehole with an effective diameter of 175 millimetres. Air was injected over a period of six weeks during 1968. The maximum quantity was 31000 cubic metres per day at 10 bars pressure.

A good quality gas - 2400 Cal/cubic metres stp - resulted. When the open cut advanced, the coal had been converted into coke and it was clear that the gas was basically a distillation product.

The next phase began in 1972 under the patronage of the US Bureau of Mines in a sub bituminous coal deposit - volatiles representing 49% daf. This was near Hanna - Wyoming.

Seam 1 is 9 metres thick at a depth of 100 to 120 metres.

A preliminary trial carried out during 1973/4 produced, for part of the time, a gas with an average calorific value of 1150 KCal/cubic metre stp at an average rate of 45,000 cubic metres stp/day. Some 55% of the energy potential of the coal consumed was recovered.

Three other projects are reported under ERDA's patronage.

- In Texas the Russians 30 x 30 metre grid borehole pattern was scheduled to be carried out in co-operation with Soviet technicians.
- In West Virginia, at Morgantown - the Bureau of Mines Research Centre - experiments are under way involving coal with a reduced volatile content. Deviated boreholes are drilled from the surface and penetrate the seam, running within it from a stretch.
- The Laurence Livermore Centre in California was charged with work on the third process. This involved the massive gasification of a deposit including a number of thin seams with intermittent dirt bands. Two holes had been drilled, at the time of the report, the seams having first been broken up with explosive charges. The results appeared to be disappointing.

Taking a further look at the extracts of the '78 allocation, it is worth noting that there are two directions for gasification.

Forward Combustion. Here the combustion front and the injected air movement are in the same directions.

Reverse Combustion. In this method the combustion front movement runs counter current to the injected air movement.

The programme listed in Table 26 really involves different beds at different depths.

Linked Wells. Reverse combustion is introduced and advantage taken of the natural permeability of the coals. A uniform bed is used.

Three other methods mentioned should be considered:

Packed Bed. This is an alternative to the linked well.

Deviated Wells - Long wall. Here, very thin seams are involved and in Eastern Coal Fields. Holes are drilled down and then along the seam.

Dipping and Drying Beds. This method is adaptable to dipping beds often at 45°.

Prior to this programme the States have no experience of gasification at medium or great depth. Two important questions remain to be answered.

- the distance required to link two boreholes
- area of gasification for each borehole

However a number of technical advances have been made of interest to other countries considering U.C.G.

- gas flow meters with interchangeable membranes.
- Pt - Rh - W - Re thermo couples for use at high temperatures.
- treatment of condensable products before analysis of the gas.
- central collection and computer data processing.

The Gulf Company is developing a low pressure gas turbine suitable for producer gas (675 K Cal/cubic metre stp). Catalytic combustion of very lean gas down to 350 K Cal/cubic metres stp is also being examined.

The key to gasification of deep deposits depends upon the linking of boreholes by hydro fracture.

Future Trends

Clearly this method - in situ gasification - should be continued. It is of particular significance to Europe as a method of recovering deep lying seams.

Heavier petroleum fractions in oil fields are already recovered through "In-situ combustion". These techniques could well be developed for tar sands, lignites and coal to which they are applicable.

Bearing in mind the cost of development of this new technology there would be considerable merit in progressing on a common front with a Europe whose interests must be directed towards recovery of coal at considerable depths.

Otherwise "in-situ" with a yield of 70% only is not very attractive.

Morocco

In the past work has been carried out by France at Djerada in Morocco. It involved a steeply sloping seam. Inlet and outlet shafts were sunk in the seam with a horizontal fire gallery connecting them along the strike. The shafts were lined with refectory material containing heat exchangers to pre-heat the inlet air. Operation was continuous over a five month period but with gas of varying quality being emitted. The scheme was abandoned due to a number of problems including surface leakage and obstruction to flow.

Czechoslovakia

There is a record of gasification of the Brown Coals of N. Bohemia. Also in 1964 use for district heating and electricity generation were proposed, but details are incomplete.

Mention is made in the records of gasification of three seams with a total thickness of 4.9 metres and a total tonnage of 108 million over 25 years; the Ash content was 45%. Electricity produced would be 220 MW and heat production at 1,000 therms per hour for the district heating scheme. That was the theory: there is no evidence as to the practice.

Sundry

Italy and Poland are also said to have carried out tests, but little record appears to remain. Canada through the Research Council of Alberta is studying alternative sources of gas, of which UCG is one. The possibility of upgrading low Btu gas to pipeline quality is of particular interest. Sites in the southern part of the province have been assessed involving a 10 foot - 3 metre seam.

Technical Requirements for In-Situ Methods

Clearly the first step to be taken is identification of suitable sites, which in Europe may involve deep-lying deposits.

The site should offer certain conditions which will make the operation easier:

- thick seams in close proximity to each other.
- coal should be permeable with impermeable surrounding rock.
- the surrounding strata must not permit leakage from the gas generator.
- large quantities of water in nearby zones must be avoided.
- the seams chosen should not be under built-up areas.
- the site should command good access to road networks.
- early tests should be carried out at depths less than 1,200 metres.

Linking of Deep Boreholes 70 to 80 Metres Apart

- permeability between a range of coals at minimum pressures will vary widely.
- permeability falls with increasing pressure. For instance, Belgian experience is that an increase in pressure from 1 to 150 bars reduces the ratio from 1 to 20 up to 1 to 300.
- coals of low initial permeability will experience a greater

reduction in permeability than those which are good, for a given increase in pressure.

Specific tests illustrating this permeability change include:

Zolder Colliery (Kempense Steen Kolen mijnen). Here water injection was carried out between 150 to 160 bar.

At 155 bar the quantity injected was 16,700 cubic metres stp over 24 hours when with a permeability of 18.3 mdarcy the proportion of water recovered amounted to about 10%.

By comparison it may be of interest to note that at the U.S.A. Hanna experimental station listed in Table the permeability of the gasified coal varied between 140 and 1400 mdarcy.

Beringen Colliery. Daily quantities between 2,500 to 3,600 cubic metres stp were injected at 120 bar. Stabilisation of the quantities injected occurred fairly quickly between 600 to 700 cubic metres stp per day over 7 days a week at pressures between 150 and 160 bar. The quantities of gas recovered at the second borehole reached 14 cubic metres stp per day, stabilising at around 4 cubic metres stp per day over 5 days a week equivalent to less than 1%.

There is a fundamental difference between water and air injection; energy stored underground by a compressible fluid is far greater than when one which is not compressible is used.

Repeat experiments carried out in southern Belgian coalfields were made for comparison purposes. Greater depths were involved, requiring a number of modifications:

- water injection under high pressure 150 to 200 bars.
- injection of air under high pressure with a counter pressure maintained at the outlet borehole to keep permeability of the seam at maximum value.
- oxygen enrichment of the air or injection of pure oxygen under high pressure to initiate counter-current combustion.

The main problems are seen to occur in setting up a pathway through a virgin deposit where hydrofracture may be required. There is a danger of vertical fissures resulting in the roof. Provided depths are sufficient this possibility may not be too detrimental to the efficiency of the gas generator. Once an initial path has been set up within the seam, linking with other boreholes will be easier to create.

Protection of the Borehole Casing

The conditions associated with contact between gases, rock and coal during underground gasification do not result in efficient heat recovery. As a result, the theoretical temperature can be an approximation only.

This must be taken into account when arriving at the maximum reaction temperature (t max = 1370°C) reached in the gasification zone, which may be calculated from the theoretical balance per kilogram of fixed carbon.

Practical experience has shown that very high temperatures will be achieved at the ends of the boreholes. The problem is to produce a borehole cooling system which will both prevent overheating of the casings as well as permitting recovery of sensible heat from the gas. The Thermodynamics Department of the University of Louvain is one of those looking into the question. Evaporation is one of the means of cooling the extremity of the borehole being examined.

ECONOMIC ASPECTS OF UCG

There is not very much evidence with regard to this important aspect.

However, for UCG to be promoted experimentally gas must be produced commercially at a cost which will make its use worthwhile.

Before arriving at estimations, certain assumptions need to be made; beginning with a unit such as 100 MW power station operating on baseload. Savings would result from the use of 1000 MW site when experience suggests that gas could be produced at between 0.3 - 0.5p/therm - cheaper than in the case of the former. This does not support the Russian claim of economy in scale.

Methods of Gasification

Two mthods only are being considered here. One method is the vertical drilling method, similar to that used in Russian commercial installations and closely approximating to the "line-drive" system which has been tested at the Laramie Energy Research Centre, Wyoming, U.S.A.

The other method involves development from an underground gallery drawn out from a shaft - commonly called the "preliminary mining method" based upon the Newman Spinney experiment - referred to under "British Progress".

These methods have been successful on an experimental or commercial scale with relevant data published. Other methods of high promise include the Belgian high pressure gasification for use with deeper coal seams which will be considered first.

Belgian High Pressure Gasification Costs

In 1965 a report on underground gasification of Austrian coal deposits was drawn up, advising that in view of the state of current technology the economic feasibility was in doubt and that research should cease.

In 1974 two researchers came to the conclusions that underground gasification was not applicable to European coal deposits on a commercial scale - Geologie en Mijnbouw, 1974. They based their conclusions on technical, environmental and economic considerations.

In early 1977 a Belgian Working Party was set up to estimate the cost price of the electricity which might be generated by a combined steam/gas plant, fired with producer gas by the INIEX underground gasification process.

One hypothetical case is considered using configuration of boreholes comprising one outlet hole surrounded by six air injection holes drilled at the angles of a hexagon of 70 metres side.

Some detailed explanation is necessary.

In this arrangement one third of the boreholes are used for gas recovery and two thirds for injection.

The first three linking pathways are at angles of 60° to 120° to each other. In this way it was believed that at least one of the three directions would be near to the direction of maximum permeability of the seam.

Certain basic assumptions were also made:

- rate of gas flow - 1200 cubic metres/stp per hour
- available energy at point of entry - K Cal per cubic metre/stp gas.

Potential heat of gas	800
Sensible heat of gas	64
Steam from cooling unit	144
	1008

Available energy at the point of entry to the station is therefore 12000 cubic metres stp/h x 1008 K Cal/ cubic metre stp = 12.1 G Cal/h or 1-73 tce.

The surface area of the hexagon of 70 metres side is 1.26 hectares, requiring a volume of coal with seam thickness of 2 metres, at 25,000 cubic metres.

365 day round the clock working, would give us an annual production from one borehole of about 15,000 tce. Therefore to achieve 1 million tce the equivalent to that of the Campine Gallery, mentioned earlier, would require 200 operational boreholes, including 67 gas outlet boreholes.

With an average life of two years for each borehole, it would be necessary to drill 100 holes per year. The 200 operational boreholes and the 100 in the process of being drilled would permanently occupy an area of 126 hectares.

It was stressed that these figures represent orders of magnitude only.

Similarly, assuming useful heat per standard cubic metre of coal gasified to be 6.5 G Cal, energy produced per borehole would be 25000 cubic metres x 6.5 G Cal/ cubic metre = 162.500 G Cal.

The total cost of the exercise is bound up mainly in the boreholes. That of a gas outlet hole (7") and two air injection holes ($5\frac{1}{2}$") will be of the order.

$$\text{B.frs } 6.6 \times 10^6 + \text{B.frs } 2 \times 5.2 \times 10^6$$

$$= \text{B.frs. } 17 \times 10^6$$

Borehole costs per G Cal represent B.frs. 105. This compares with B.frs. 220 per G Cal for raw imported coal leaving considerable room for inflation to raise the cost of depreciation, piping, labour and services.

At the same time, labour costs associated with drilling 100 boreholes per year are much smaller than those employed in producing 1 million tons of coal per year. Labour costs have a significant bearing but gasification brings the added bonus of the removal of men from underground.

Certain assumptions have had to be made. These are listed in Table 27.

TABLE 27 General Assumption used in Vertical Drilling Method

		Vertical drilling method	Preliminary mining method
i)	Vertical bore-holes	100ft (30m) apart (at corners of squares)	75ft (23m) apart (in rows 900ft (275m) apart)
ii)	Galleries	-	300ft (900m) long
iii)	Horizontal holes	-	75ft (23m) apart 900ft (275m) long
iv)	Barrier width	-	10 x seam thickness
v)	Proportion of coal burnt	85%	85%
vi)	Proportion of heat gained	60%	60%
vii)	Overall efficiency (v x vi)	51%	51%
viii)	Gas calorific value seams less than 6ft (2m) thick	80 Btu/cubic feet (3MJ/cubic metres)	80 Btu/cubic feet (3MJ/cubic metres)
	seams 6ft (2m) thick or more	100 Btu/cubic feet (3.7 MJ/cubic metres)	100 Btu/cubic feet (3.7 MJ/cubic metres)
ix)	Leakage rate	4%	4%
x)	Operating pressure	30 psig (2 bar)	30 psig (2 bar)
xi)	Reactant	Air	Air
xii	Average reactant rate	100 cubic feet/sec (2.83 cubic metres/sec)	100 cubic feet/sec (2.83 cubic metres/sec)
xiii)	Linkage air rate	66,000 cubic feet/foot (6,124 cubic metres/metre)	

The layout in its simplest form is shown in Fig. 35.

It consists of a grid of boreholes with an internal diameter of 4m (0.1 metres) lined with steel casings and placed 30 metres apart. The borehole spacing is a major cost factor, increased spacing reducing costs, although not adopted in experimental work. The boreholes are linked in the coal by high pressure air, 1 psig being considered necessary for each foot of overburden at a rate of 66,000 cubic feet for each foot of linkage through the coal - 6,124 cubic metres/m. This rate is higher than that used at Hanna to allow for the lower permeability of British Coals. Air is used for the reaction at 30 psig or 2 bars, at a rate of 100 cubic feet per second - 2.83 cubic metres per second with a 4% leakage loss.

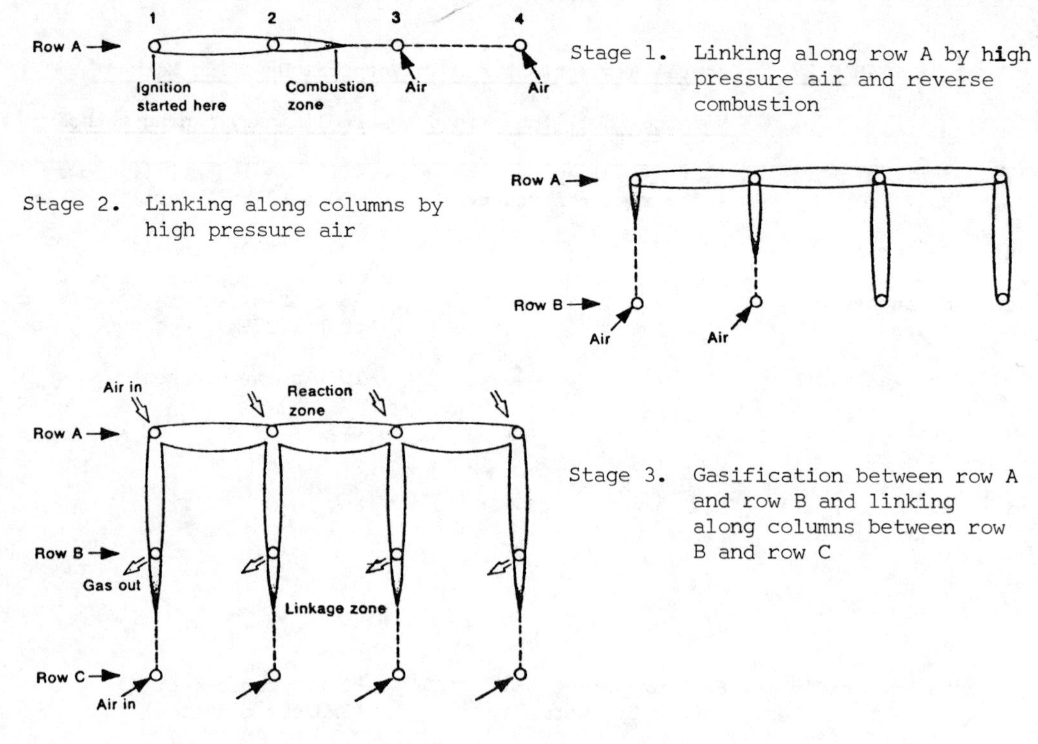

Fig. 35 The vertical drilling method showing the three stages of development

The oxygen content of the air reacts with the coal to produce 1.2 volumes of gas for each volume of air supplied; the calorific value of the resulting gas, depending upon the thickness of the seam. It was assumed that gas at 80 Btu/cubic foot would be obtained from seams less than 2 metres thick with gas at 100 Btu/cubic foot from seams in excess of 2 metres. This is based upon Russian experience with overall gasification efficiency taken at 51%, burning 85% of the coal and converting 60% of its heat into gas.

Using these figures, a borehole in a 4 foot - 1.2 meteres - seam would enable 1,500 tons of coal to be gasified producing 200,000 therms of gas.

Mining requirements. A shaft and gallery are necessary, to permit men and machinery to enter, with a concrete lining and man-riding facilities. The gallery in this instance would be 3,000 feet - 900 metres - long. Air conditions have already been outlined.

It has been assumed that ideal conditions with regard to disturbances, etc. will prevail and that there will be standard permeability.

Equipment and Labour. Costs have been assumed to be those prevailing in British Mining during 1975, which because it is a nationalised industry do not generally vary so far as wages are concerned. Construction has been assumed on a round the clock basis, but savings would result if day shifts only were introduced, since drilling accounts for a large proportion of costs.

Cost of gas. The coal to be used possesses the following characteristics:

Calorific value	- 12,100 Btu per lb.
Ash	- 10%
Moisture	- 5%
Sulphur	- 1%
Mineral matter free	- 14,500 Btu per lb.

Costs in pence per therm for diluted gas include cooling and cleaning but not sulphur removal.

Gas costs using the vertical drilling method can be seen in Table 28 where depths up to 3000 ft. - 900 metres - have been used.

TABLE 28 Examples of Gas Costs - Vertical Drilling Methods

Seam thickness		Seam Depth		Gas Cost
feet	metres	feet	metres	per therm
6	1.8	500	150	7.7p
5	1.5	500	150	8.6p
5	1.5	1000	300	10.9p
5	1.5	2000	600	15.4p
4	1.2	1000	300	12.4p
4	1.2	2000	600	18.0p
4	1.2	3000	900	23.5p

Costs shown in Table 28 range between 7 and 24p.

Similar examples are given in Table 29 for the preliminary mining method.

TABLE 29 Examples of Unit Gas Costs using the Preliminary Mining Method

Seam thickness		Seam Depth		Gas Cost
feet	metres	feet	metres	per therm
6	1.8	500	150	6.8p
5	1.5	500	150	7.5p
5	1.5	1000	300	8.3p
5	1.5	2000	600	9.7p

Seam thickness feet	Seam thickness metres	Seam Depth feet	Seam Depth metres	Gas Cost per therm
4	1.2	1000	300	9.2p
4	1.2	2000	600	10.9p
4	1.2	3000	900	12.8p

It can be seen that gas costs by this method range between 6 and 13p.

Comparing the two methods it is clear that preliminary mining reduces the cost of gas very considerably as compared with development totally from the surface. The latter must be used at depths outside the range of normal mining.

Adjustment of the distance between boreholes could reduce the cost of gas from the vertical drilling method making it more competitive with preliminary mining, but considerably more work needs to be done before this becomes fact.

High Ash Coals. The change in cost per therm remains fairly constant up to an Ash content of 25% as shown in Table

TABLE 30 The Effect of Ash Content upon Cost per therm

Method	10% Ash 12,100 Btu/lb.	25% Ash 9,710 Btu/lb.	50% Ash 5,730 Btu/lb.
Vertical drilling	12.4p	13.0p	15.9p
Preliminary mining	9.2p	9.5p	11.3p

Conditions attached to figures shown in Table 30 assume - 4ft thick seam 1000 ft. deep (1.2 metres thick seam 300 metres deep).

The rise in cost between 25% and 50% Ash due to the reduction in the amount of carbon in the seam, each borehole producing fewer therms.

Transport costs. UCG gas being of low calorific value is expensive to transport - 10 miles being about the limit as can be seen in Table 31.

TABLE 31 Transport Costs for Low Btu Gas at 20°C per therm

Distance	1 mile (1.6 Km)	10 miles (16 Km)	50 miles (80 Km)
Transport Cost	0.04 - 0.05p	0.7 - 2.8p	6.0 - 6.5p

The higher figure in each group involves a provision for pipe laying across built-up areas.

Cost Reduction

There are some seven factors which can affect the cost of the process. These are:

Borehole Spacing. The most likely change would be to increase the distance between boreholes as applied to the vertical drilling method. Similarly, larger galleries and horizontal boreholes would affect costs in the preliminary mining method, beneficially. Table 32 indicates the cost reduction to be expected from increased distance between boreholes.

TABLE 32 Economic Effect of Changes in Borehole Spacing in the Vertical Drilling Method

Distance between boreholes in ft.	60	100	200	400
Cost/therm of gas	27.8p	12.4p	7.3p	5.9p

Conditions attaching to Table again assume a 4 ft. (1.2 metres) thick seam, 1000 ft. (300 metres) deep.

It can be seen that the cost falls dramatically at 200 ft. gaps and even more so at 400 ft., all other things being equal.

This is supported by experience at Laramie where borehole spacing of 200 ft. and 400 ft. have been included.

Gasification Efficiency. This is a way of expressing the amount of heat which can be recovered from the coal. It is reflected in the number of panels required and in turn the area of the site. Greater efficiency will result in less drilling etc.

Table 33 indicates the effect of changes in efficiency.

TABLE 33 Effect of Overall Efficiency on Unit Gas Costs per therm

	Overall Efficiency		
Method	40%	51%	65%
Vertical drilling	14.3	12.4	11.0
Preliminary mining	8.6	9.2	7.5

It should be noted that efficiencies lower than 40% have been recorded while 51% would appear to be a reasonably attainable level. Also the difference in cost pattern resulting from the two methods.

Pressure. If an air supply at the rate of 100 cubic feet/sec - 2.8 cubic metres -

and with between 14 and 18 panels operating together, were to be reduced proportionately, more panels would be needed. This would require increased quantities of piping and monitoring equipment. The effect of the changes are shown in Table 34.

TABLE 34 Effect of Operating Pressure on Unit Gas Costs per therm

Method	20 psig (1.4 bar)	30 psig (2 bar)	300 psig (20 bar)
Vertical drilling	12.1p	12.4p	15.7p
Preliminary mining	8.8p	9.2p	12.4p

Conditions assumed here are as before. The increase in cost is less for a given change in pressure than might have been expected.

Calorific value of gas. A rise in calorific value of the gas above 100 Btu/cubic foot would require a smaller volume of gas to produce the same amount of heat. Less air would be needed. In turn smaller compressors and reduced power consumption would follow and therefore lower costs.

The effect of these changes is shown in Table 35.

TABLE 35 Effect of Gas Calorific Value on Costs

	Gas Calorific Value - Therms		
Method	70 Btu/ft^3	80 Btu/ft^3	160 Btu/ft^3
Vertical drilling	12.6	12.4	11.6
Preliminary mining	9.3	9.2	8.3

The gas values range between the attainable and the unlikely, 160 Btu/cubic foot being very difficult to reach. It would probably be uneconomic even were it to be achieved by upgrading.

Gasification of more than one seam. No experience of gasification of larger seams has so far occurred, the exercise to date assuming a single seam only, dirt bands and all. The introduction of more than one seam should reduce costs considerably in that each borehole would lead to a larger volume of coal.

Use of oxygen. This method provides a means of producing a higher calorific value gas and around 250 Btu/cubic foot - against 80-100 Btu/cubic foot when air is used. As a result, nitrogen-free gas is a better feed stock for upgrading to SNG, making it cheaper to transport. There would however be an increase in cost of between 13-14p/therm than were air to be used. Oxygen derived from the nuclear splitting of water together with Hydrogen for upgrading could in the future, be more attractive. For a 100 MW site some 2500 tons of oxygen per day would be required - a very considerable quantity by industrial standards.

Comparisons

Clearly UCG has certain advantages as well as disadvantages. The number of disadvantages at least in numerical terms, outweighing the former. A more subjective judgement can be reached by comparing them.

Advantages

- It provides a potential method of exploitation of reserves which are not amenable to conventional mining.
- It provides an indigenous gas source.
- The gas is a potential chemical feedstock.
- Energy is produced in fluid form giving at least some potential for flexibility in use.
- Environmental effects would be temporary.
- Ash is left underground and less tipping dirt is produced compared with conventional mining.
- Labour intensity is low. Either no men have to work underground or if they do they are few in number and removed from the points of greatest danger.

Disadvantages

- The extraction of reserves is slightly less than that by deep mining. The energy balance is also less favourable than that of deep mining.
- Costs are at the upper end of the range of existing energy costs. The cost per therm is always higher than that for natural gas at source and opencast mining.
- Capital input is rather greater than that for deep mining.
- The gas produced is dirty, has a low calorific value, and is variable in quality.
- The cost of making SNG from the gas would be higher than that of importing liquid natural gas.
- The gas is costly to pipe much over 10 miles (16 Km) and to store.
- Temporary visual damage to the environment would be caused by drill rigs, headgear and piping. Topsoil could be damaged by movement of heavy machinery. There is a temporary requirement for large areas of land which in some countries could be restrictive.

- There would be noise from drilling.
- New skills would have to be taught. For a time these skills are likely to be in short supply.

The gasification of coal is a vast field with much research required in the case of in-situ and development across the board.

The processes are known, but there is a considerable difference between a small gasifier capable of sustaining an isolated plant and an arrangement capable of making use of the vast transmission and distribution networks which already exist, even in Europe.

In this Chapter an attempt has been made to review the work being carried out in the major industrial countries together with the identification of site locations and organic actions involved.

The British Westfield project with its very considerable U.S. backing is clearly of major significance while the U.S. programme of over $100 million dollars provided by the Department of Energy in 1978 alone, is a clear indication of the need seen in that country.

The projects must be made to succeed; the Chemical Industry is highly dependent upon a feedstock which long term, coal alone appears to be capable of providing.

Chapter 10

CONVERSION AND PETROCHEMICALS

The liquefaction of coal provides an alternative route to gasification in the production of a feedstock for the Chemical Industry. From 100 tons of coal about 20 tons of petrol and a range of aromatics yielding about 40 tons plus tar residues result. But as the United States Committee on Appropriations of the House of Representatives heard when evidence was taken in April 1978, Liquefaction is of interest basically because it is the only route to producing a liquid fuel for transport.

However, there are a number of countries, at the very least taking an interest in the subject.

Table 36 indicates those countries which have made provision for involvement in liquefaction in their expenditure programmes.

TABLE 36 Comparison of Expenditure for involvement in Liquefaction

Millions - Units of Account

Year:	1974	1975	1976	Involvement
Belgium	-	.04	.04	General Coal
Britain	0.67	1.63	?	See text
France	.05	.05	.07	
Germany	19.27	3.11	3.00	See text
Italy	-	-	-	Feasibility
Netherlands	-	-	.05	Assessments

In addition South Africa and the United States have considerable involvement. Both are reviewed later.

Table 36 has been drawn up in common currency terms so that a comparison as to expenditure can be made. Belgium, France, Italy and Netherlands are merely flirting with the subject.

Britain

Here interests are different to those of The States; Belgium seeks:

- High grade products for electrodes and carbon fibres.
- Development processes to make synthetic hydrocarbons as a feedstock for the petrochemical industry.

There are two methods involved both being financed by the European Coal & Steel Community grants.

The first process is essentially one of liquefaction, using liquid solvents. This is shown diagrammatically in Fig. 36.

The layout is indicative of a number of fundamentals:

- The need to generate sufficient oil during hydrogenation of a quality which can be recycled.
- To reduce the soluble residues to a minimum.
- The vital role of efficient catalysts.
- The necessity to obtain hydrogen from either the residue or the coal itself.

Liquid Solvent Extraction

This process involves the mixing of coal with oil in the ratio of 1 to 4 at a temperature which results in the removal of moisture. After this, the slurry is passed through a preheater at 400°C followed by digestion under pressure, and then filtered at 200°C.

In this way, some 90% of the coal becomes dissolved with a resultant ash level of less than 0.1%. The resultant extract can be used for hydrogenation or for carbon products, including carbon fibres. Figure 36 indicates the general process for making carbon fibre and calcined coal from coal extract.

Supercritical Gas Method

In this method, instead of conventional liquid, the gas is compressed beyond its critical pressure. With supercritical gases vaporisation can be induced much more readily in a range of substances. Choice of a gas is normally governed by the proximity of the critical temperature to that of extraction. Coal breaks down at around 400°C, when liquids are formed but do not distil, while at higher levels they decompose to coke, gas and liquor. Extraction gases for coal require critical temperatures at this level - that of coal tar fractions or naptha. However since water acts as a solvent, predrying becomes unnecessary. None of the ash dissolves, remaining behind with the coal. The extract therefore is basically ash and is solvent free.

This method results in a lower molecular weight and higher hydrogen content than

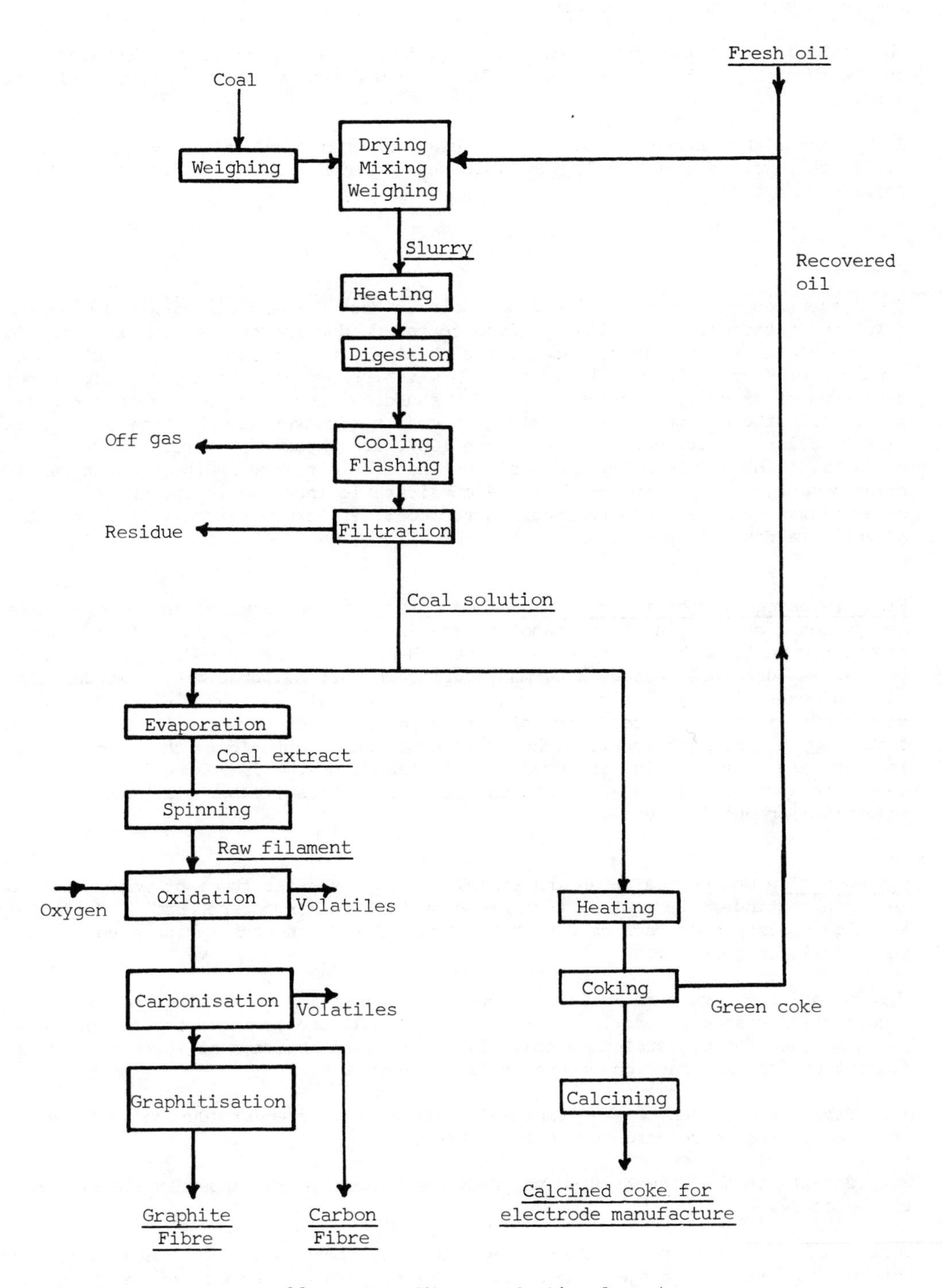

Fig. 36 Carbon fibre production layout

those produced via the liquid solvent route. They are more readily converted to fuel oils and chemical feedstock.

The yield is lower than by liquid solvents, but on the other hand it is higher than with carbonisation processes. This is even more marked when poor coals are used.

A char forms the extraction residue not dissimilar in size to the original coal for which power stations employing fluidised combustion and gasification would provide an outlet.

Germany

The conversion of coal into liquid products in West Germany represents a major development target. Priority is given to the production of chemical raw materials because of the immense importance of the chemical industry to the national economy. A number of studies in this field have, however, shown that at present day prices this type of process is uneconomic. The source of coal does play a significant part, while the use of lignite instead of coal for conversion to chemical feed is said to raise the price by between 30 to 100%, as compared to petroleum derived products. This could alter and parity of cost be obtained if the present research programme proves to be successful. A considerable increase in the price of oil or a massive subsidy would be required for coal, if German pit coal is to be competitive for this purpose.

Fischer-Tropsch Synthesis process. This method, being examined in Germany, uses as synthesis gas similar to methanol synthesis gas, but as a result of different reaction conditions hydrocarbons and other chemicals are produced. The general process was originally used in Germany during the Second World War. It was also used in Texas with synthesis gas produced from natural gas, but was unsuccessful on grounds of cost. Success in this field demands that advances in catalyst technology are brought about, although feed stocks acceptable to the chemical industry are already being produced, in the laboratory. Increased life of the catalysts must now be the aim, with the samll scale results being repeated on an industrial production level.

Hydrogenation was proved on an industrial scale just after the last war. This has led to a continuous laboratory test known as the "IG Farbon Process". Good yields and high quality hydrocarbons have been obtained and a demonstration plant on a tonne scale, completed.

Hydrogenation offers a reliable and low-cost process for converting coal into a liquid of increased calorific value "in situ". Remote or inaccessible deposits could be used for conversion to coal oil and then transported at low cost. The funding of the liquefaction programme is shown in Table 37.

From Table 37 it can be seen that while actual disbursements during 1976 were low, the average percentage support was high.

Full details of the liquefaction programme are shown in the Appendix at the end of the study.

Details of the West German programme are the most comprehensive. They contain not only the title of the project, but also a description and details of the participants.

TABLE 37 Project Breakdown of Costs for Liquefaction Projects begun in 1976

Millions DM	
Total Cost	96.7
Total Support Grant	81.7
Disbursements during 1976	5.4
Average percentage support	83.0

South Africa

South Africa is in the process of tripling production of gaseous and liquid hydrocarbons. The new complex known as Sasol II cost in excess of $1 billion and was constructed in South Africa's Transvaal. It was designed to process coal into gasoline and diesel fuels, ethylene, ammonia, sulphur and a range of coal-tar derivatives. The plant will be similar to the earlier version Sason I. It should be capable of supplying some 40% of South Africa's needs for motor fuel.

Coal preparation. Three beds are provided, one above the other. The lowest bed has a mining height of about ten feet. It will be mined first. Continuous mining takes place in conjunction with room and pillar, longwall and shortwall techniques. Coal is transported on conveyors direct to the preparation plant from which it passes to the gasifiers and boiler plant.

Steam only is used for the process. This demands strict sizing so that gasification is as effective as possible. Fines tend to be carried through the gasifier, emerging with the raw gas which creates problems with the product as well as loss of yield. Minimum size of coal is around 6 mm.

Gasification. A Stretford plant for sulphur recovery forms part of the complex Sasol II recovers ammonia as anhydrous ammonia, while air and water are used for cooling, the system depending mainly on climatic conditions, availability of water, quality of coal and the way in which the gasifier is operated to meet downstream needs and general steam balance.

Fischer-Tropsch. The object is to produce more fuels and therefore fluid bed reactors have been installed. Oil workup has been designed to use hydrocarbon liquids which come from the fluid bed Fischer-Tropsch synthesis reactors. This hydrocarbon is liquid, clean olefinic in properties as opposed to the type of crude used in refineries. The refining methods produce high grade motor and diesel fuels.

The chemical workup employs precise separation of oxygenated compounds to produce higher quality alcohols and ketones. The aqueous oxygenated chemical solutions are first fractionated to acids and non acids, the former, together with the water being discharged for treatment, being biodegradable. Non acid chemicals are then separated into carbonyls and alcohols later converted to ketones and alcohols.

United States

The main objectives of the programme are:

- the conversion of domestic coals into chemical feed distillate,heating oil, gasoline and boiler fuels
- commercial second generation liquefaction
- near commercial scale plants before 2000 A.D.
- commercial plants after 2000 A.D.

The Federal role is to accelerate development through cost sharing with industry, as well as providing supporting research incentives for commercial use. One prime property of liquefaction is that it is seen as a means of reducing sulphur for liquid feed and fuel purposes.

The 1978 United States Liquefaction Programme is shown in Table 38.

TABLE 38 U.S. Liquefaction Programme & Funding - 1978

Project	Product	Funding (000's)	Status
1. H-Coal Pilot Plant	Syncrude & Boiler Fuel	$26,000	Completed
2. Synthoil PDU-Perc	Boiler Fuel	11,000	Initial operations
3. SRC Pilot Plant	Solid Fuel	15,000	Operation for data base
4. Donor Solvent Pilot Plant	Syncrude	30,300	Detailed design
5. Cresap Facility	Equipment Tests	12,500	Operations
6. Coalcon Test Bed (Site undecided)	Clean Boiler	10,000	Process evaluation
		$104,000	

It should be remembered that the figures shown are the 1978 Budget Allocation reduced by around 10% from the original request.

Some comment with regard to the processes may be useful.

Synthoil Process

This process backed by the United States Bureau of Mines is a single stage catalytic process employing a fixed bed. The catalyst used is cobalt molybdate, supported on alumina.

Gas, oil and coal are injected at a high flow rate to ensure turbulence and deter the catalyst bed from clogging. This is assisted by the use of finely ground coal and operating conditions with pressures of 2,000 to 4,000 psi at 850°F, with a passage time of two minutes only. Oil used in the slurry forms part of the product. A considerable range of coals have been used. A boiler fuel results from the process.

H-Coal pilot plant. This project is now completed following satisfactory results on a bench-scale and process development work. This was followed by the catalytic liquefaction method used at Catletburg, Kentucky. The process was developed by Hydrocarbon Research Inc. for conversion of coal to hydrocarbon liquids and rich gas. It is capable of up to 95% conversion of organic matter in the coal, to gases and liquids.

A wide range of coals have been used during the development stages from eastern, mid-west and western bituminous coals, to western sub-bituminous coals, to lignites from Texas and North Dakota and Brown Coal from Australia. The main work has been carried out using Illinois No. 6 and Wyodak coal, but no coal tried has failed to be converted satisfactorily by the H-coal process.

Coal is slurried with recycle oil and fed with hydrogen to the bottom of the H-coal reactor system. Here the reactants come into contact with an ebullited catalyst bed. Unconverted coal and ash pass up through the catalyst zone and are entrained in the liquid leaving the reactor. The catalyst is maintained in an expanded state by internal recycle of catalyst-free liquid from the top of the reactor back to the bottom. Here, the liquid mixes with the incoming feed.

Being a back-mixed reaction system, with catalyst in random motion, there are a number of advantages over a fixed-bed, namely:

- The reaction zone which is basically isothermal can be maintained at optimum temperature, relative to the product, by controlling the temperatures of the feed streams. They will probably remain up to 200°C lower than the reaction temperature. The exothermal heat of reaction is used directly at 100% thermal efficiency.

- Catalyst can be changed maintaining the unit in operation.

- Unconverted coal and ash are able to pass through the catalyst bed freely.

- The Catalyst is selective for both sulphur conversion and upgrading the liquid, with minimum conversion of liquids to gases requiring a higher hydrogen/carbon ratio.

The versatility of the process for producing a wide range of liquid and gaseous products from a range of coals has been demonstrated. So far as liquefaction processes go, it is probably without peer. Its particular virtues are a capability to comply with the strict pollution restrictions which will increase with time.

Solvent refined coal process - SRC. This process is funded by the Office of Coal Research and carried out by the Chemical and Petroleum Refining Engineering Department of the Colorado School of Mines. The study was originally set up as a three phase project scheduled for three years. The main object was to study the desulphurisation of coal by this process.

The first phase considered five operating variables:

- Temperatures: 325°C and 400°C.
- Partial pressure of hydrogen 600 and 1,200 psi.
- Solvent-to-coal ratio 2:1 and 3.5:1
- Solvent type: anthracine oil and tetralin
- Reaction time: 7.5 min. and 15 min.

Temperature, pressure and solvent-to-coal ratio were found to be the most important factors in the removal of sulphur, while anthracine was the superior solvent. This information was used in the second phase.

Batch solvent refining of coal char reduced one third of the sulphur content with a set of conditions including:

anthracine oil solvent

pressure at 2,000 psi

high temperature at 425°C

high solvent-to-char ratio 5:1

The present pilot plant project at Tacoma, Washington makes use of this slurry recycle to produce increased hydrogenation and provide test products.

Donor Solvent Extraction. The production of distillate "syncrude product" in the primary liquefaction stage results from this process. The carefully tailored donor solvents used are recovered later, by distillation to result in solid liquid separation. The recycle solvent, freed from solids, is rehydrogenated in a conventional fixed bed catalytic reactor. This plant is at Baytown, Texas and has now entered the detailed design stage.

Cresap Facility. This plant was set up at Cresap, West Virginia, with the original intention of developing the technology for conversion of high sulphur caking coals into low sulphur distillates using coal extraction, as a first step, followed by catalytic hydrogenation.

Four stages were involved:

- Coal extraction at 750°F with 80% yield, on moisture and ash free feed coal.
- Solids separation at 600°F and 150 psi with 90% recovery of extract at 1% ash or less.
- Carbonisation of the residue at 800 to 900°F, at low pressure for solvent recovery, the char being suitable for hydrogen production.
- Vacuum distillation at below 650°F fluid temperature with short residence time.

More recent work, that featured in the 1978 budget allocation, is said not to be proceeding as well as the earlier efforts. This should be confirmed.

Liquefaction of lignite. A non-catalytic process has been under development to produce low sulphur fuel oil through the reaction of synthesis gas and steam with poor quality coal, such as North Dakota lignite.

A 30% lignite in anthracine oil slurry was fed together with gas and water into the bottom of a reactor and then on to a high pressure receiver where gas-liquid separation occurred.

A benzene-soluble material made from lignite emerged. The work has been proceeding using coal slurries in processed derived oil to obtain a product oil from a variety of lignites and bituminous coals so that data for scale up may result.

CHEMICALS FROM COAL

Mention has been made on a number of occasions of the way in which coal will fill the gap left by oil and natural gas when supply ceases to meet demand. This will happen, long before reserves become exhausted.

There are two routes to a chemical feedstock:

1) Through gasification of coal either on the surface in gasifiers followed by methanation or upgrading through the removal of nitrogen to medium Btu gas or UCG and similar improvement processes.

2) By liquefaction and subsequent fractionation.

Before examining these processes it would be useful to consider the composition of coal.

In chapter 2, a brief review of the way in which coal was formed took place. Increased severity of conditions and the period of formation increases the rank or quality of the coal. This is often accompanied by a reduction in hydrogen and oxygen content. The real determination of the rank of a coal is obtained from an analysis, its calorific value and its tendancy to agglomerate. The analysis and the properties usually measured, were listed in chapter 2.

The hydrogen content of the coal is particularly important in this context because the higher the level, the less the hydrogen needed to convert the coal to liquid or gaseous fuels. The amount of hydrogen measured on a moisture and ash free basis (MAF) is around 5 to 7% for wood, peat, lignite and high-volatile bituminous coals falling progressively in low-volatile bituminous and anthracite coals to 2% or less. By comparison with petroleum and oil shale, coal has a much lower hydrogen content, bituminous coals enjoying half that of oil only, while containing more than ten times the oxygen present in oil.

Coal Combustion

When burned with an excess of oxygen the only bi-products of potential use to be obtained from coal are sulphuric acid or elemental sulphur. Virtually all the sulphur in the coal is converted to sulphur dioxide and will probably require removal either by scrubbing the gases or through one of the processes described earlier in which the sulphur remains in the ash.

Similarly, the nitrogen in the coal is converted to elemental nitrogen and oxides of nitrogen, the latter forming a potential source of nitric acid for which there is no practical form of recovery.

The carbon and hydrogen in the coal are converted to carbon dioxide and water.

Coal Carbonisation

This process amounts to the distillation of coal to a point of destruction in the absence of oxygen. Another name is coking, forming the basis of organic chemical production since it was first used probably by the Germans during the last century.

A brief list of the products of carbonisation together with some of the derivatives gives an idea of the vital role which coal has to play in the future prosperity of the chemical industry. Some of the derivatives will require elements from sources outside the coal, as can be seen from this theoretical family tree of coal derivatives.

Gas

- Hydrogen sulphide
- Hydrogen cyanide
- Hydrogen
- Carbon monoxide
  - Methanol
  - Carbon dioxide
    - Urea (requires ammonia)

Gas liquor

- Pyridine tar bases
  - Pyridine
  - Picolines
  - Ammonia liquor
  - Ammonium sulphate

Light Oils

- Benzene
- Toluene
- Xylenes
- Carbon disulphide
- Cyclopentadiene
  - Resins
- Heavy naphthas
  - Resins
  - Paints and varnishes
  - Rubber chemicals

Tar

- Carbolic oil
  - Crude tar acids
    - Phenols
    - Cresols
    - Xylenols
  - Neutral oil
    - Tar bases
    - Naphthalene
- Light creosote oil
- Heavy creosote oil
  - Heavy tar bases
    - Quinolines
      - Dyes
      - Pharmaceuticals
    - Quinaldine
- Rubber softeners
- Anthracene
  - Anthraquinone
- Phenanthrene
- Carbazole
  - Resins
  - Vat dyes
- Refined tar
  - Pharmaceuticals
  - Tar products
- Pitch

Coke

- Metallurgical coke
- Graphite
- Briquets
- Water gas
- Calcium carbide
  - Calcium cyanamide
  - Acetylene

The way in which the carbon, hydrogen, oxygen, sulphur and nitrogen are found within the coals, coupled with the conditions under which processing occurs will govern the yield, according to the way in which they can or may be found in the various products including gas liquids and coke. Ammonia, hydrogen sulphide, hydrogen, carbon monoxide and a range of aromatics all yield more readily as coking temperature increases.

If attention is focussed on methane, then a theoretical situation arises as shown in the flow diagram in Fig. 37.

A fuller list of petroleum feedstocks and some of their derivatives appear in Appendix

However a series of products from "gas" can be expected to follow this pattern.

Hydrogen

- Cyclohexane (from benzene)
  - Plastics
  - Fibres
- Ammonia (from nitrogen)
  - Urea (with carbon dioxide)
    - Fertilizers
    - Resins
  - Nitric acid
  - Fertilizers

Methanol

- Formaldehyde
  - Resins

Carbon monoxide

- Formamide
- Sodium formate
- Carbon dioxide (for urea)
- Phosgene (with chlorine)
  - Polycarbonate resins
  - Isocyanates
    - Polyurethanes
  - Carbamates

Carbon dioxide

OXO aldehydes

- Alcohols

Fischer-Tropsch products

- Olefins
  - Polyolefins
  - Alcohols
- Alcohols
  - Methanol
  - Ethanol
- Ketones
  - Acetone
  - Methly ethyl ketone
- Organic acids
- Fuels and waxes

Potential for Chemicals

While no-one can by now be in doubt as to the importance of coal and even the role which it will play in tomorrow's technology, it may not be quite so apparent how it can influence our lives through products manufactured by the Chemical Industry.

Ammonia. This is a very large volume product used for fertiliser, by direct injection as in Denmark or else in the form of ammonium sulphate or nitrate. Large quantities of Ammonia are used in polymers, mainly acrylonitrile and certain amines.

Ethylene. This is converted to polyethylene either for plastics fibres or for chlorination to vinyl chloride and styrene. It may be oxidised to ethylene oxide particularly for use in ethylene glycoal-anti freeze. There are may other diverse

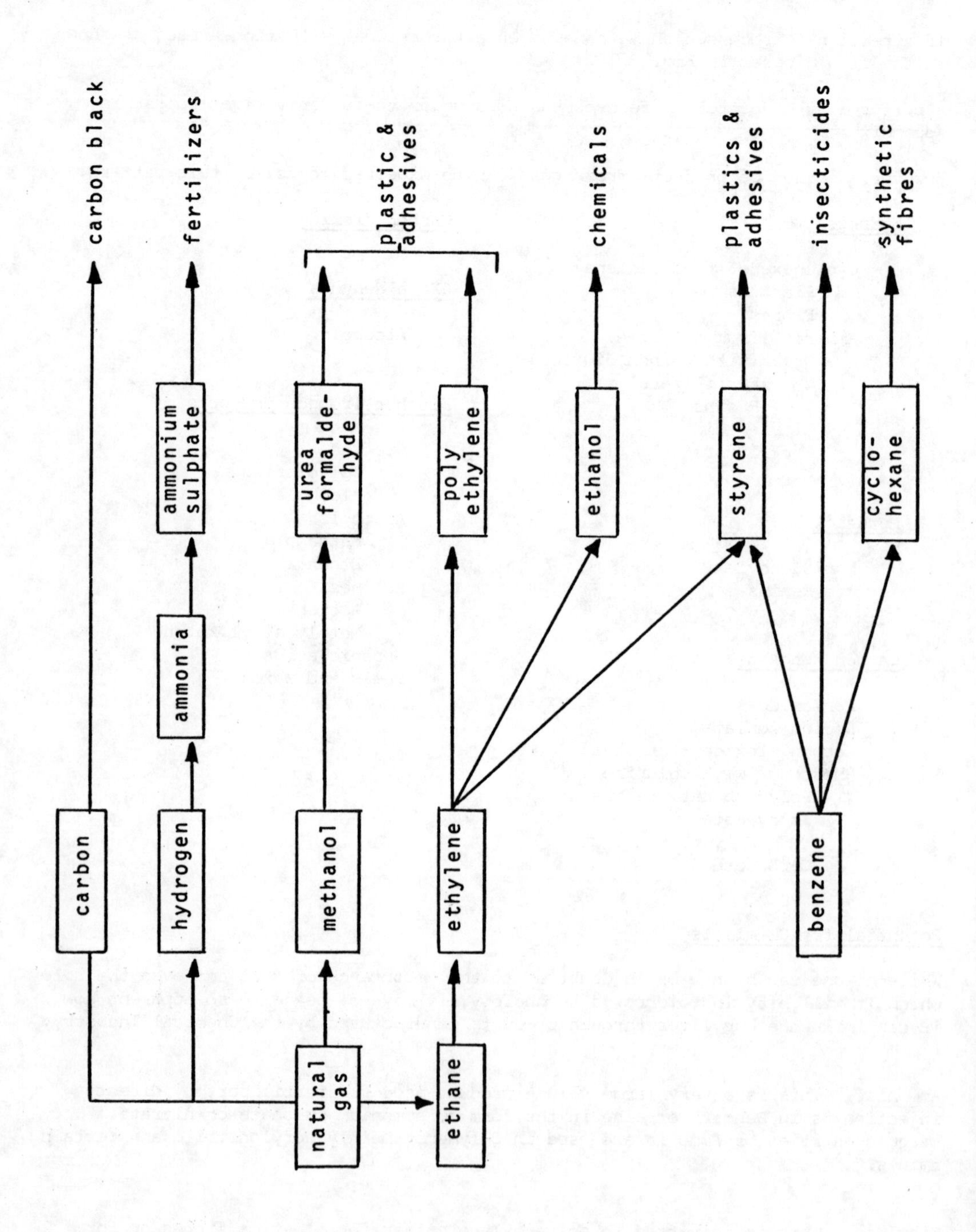

Fig. 37 Natural Gas - Chemical Feedstock

uses.

Propylene. This product may be polymerised directly but for more gas for non-polymer uses than in the case of ethylene, one of the largest uses being isopropanol. Propylene is used to make acrylonitrile, propylene oxide for purposes parallel with ethylene oxide as well as an intermediate for acetone and phenol used in the plastics industry.

Butadiene. This is used almost exclusively for polymers. Benzene is the raw material for styrene, phenol cyclohexane and maleic anhydridge. Synthetic detergents, aniline, nitrobenzene and chlorinated aromatics are also important uses. Toluene arises in pure form during the production of benzene and xylenes only to be blended into petroleum as a high octane component.

Orthoxylene is used for phthalic anhydridge running parallel with naphthalene for use in polyester resins, while Paraxylene is used in the production of terephalic acid or dimethyl terephthalate for use in polyester fibres.

Methanol. This intermediate goes into formaldehyde mainly for plastics, although it is also used for methyl esters, amines, solvents and a number of other products. More recently, ICI have developed a method of obtaining protein suitable for animal feed from Methane, via Methanol. The layout for the plant is shown in Fig. 38.

The process, which will begin at ICI's Billingham complex in Britain by 1980, is based upon a continuous fermentation of methanol using selected micro-organisms. Considerable amounts of air are supplied to the fermenter for respiration of the culture.

In all, water, air methanol, ammonia and minor inorganic nutrients are fed to the fermenter, from which a three per cent dry weight suspension of cells is removed continuously.

Although designed primarily for animal feed it will release traditional crops for human consumption. With a methanol-protein ratio of 2:1 about one million tons of additional methanol will be needed, rising to two million tons by 1985 and possibly five million tons by 1990. The value of methanol being exported in the form of protein is not only a more convenient form, but enjoys a very considerable added value as compared with methanol.

Methanol from Coal

The use of methanol in the important ICI process which produces protein has already been mentioned. More recently, tests have confirmed that methanol is a satisfactory nonpolluting boiler fuel and already the two largest power companies in Southern California in the United States have been recommended to burn methanol, instead of fuel oil, during the 45 days in mid-summer when air pollution becomes unacceptable.

The use of methanol as a substitute for petrol or at least as an additive highlights the answers to a number of questions :-

- methanol now leads the field as a petrol supplement

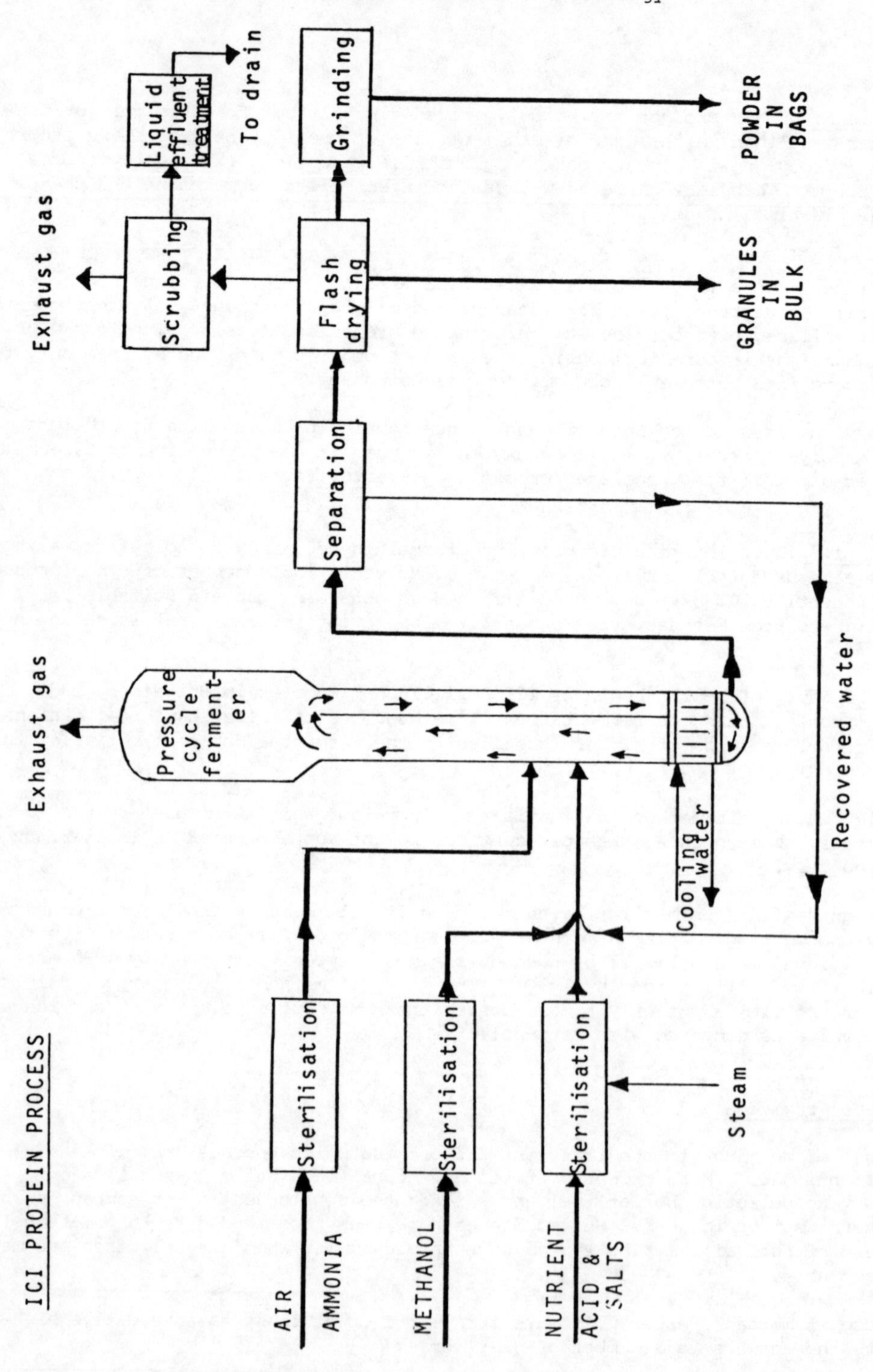

Fig. 38 Protein from Methanol plant layout

- methanol can be used as a 15% blend with petrol or gasoline without any major changes to the carburetor or engine. Certain plastic components in the fuel system will need to be replaced by metal.

- Volkswagen carried out tests throughout Germany over a two year period using a fleet of 45 vehicles under all types of driving conditions. These have shown that many of the potential problems which had previously appeared to limit the use of methanol could now be discounted.

- Significant reduction in the emission of carbon monoxide, unburnt hydrocarbons and carcinogenic polynuclear aromatics results.

- engine efficiency was consistently better - 3% - than when petrol was used.

In the early days, these uses will result in a major traffic in methanol from oil and gas fields in the Middle East, but later coal based medium - Btu plants will be used. This is already happening in the United States, where all additional production of methanol will come from coal via the Winkler process.

Future Partners for Coal

Clearly the need for polymers will increase rapidly over the years ahead. As current natural gas and its liquids become reduced, so they must be replaced with heavier liquids for use as a feedstock. This would identify petrochemicals with the petroleum industry, particularly in Europe where Naphtha is dominant over Natural Gas for a number of purposes.

Most producers of ethylene have down stream needs for propylene and butadiene. In the United States the converse is the case, making ethane a high suitable feed.

However, with coal being used as a feed, ammonia or methanol are basically the only products which can be made and therefore a liaison with the energy industry would be extremely likely. But there are other factors to be considered.

Over the years, the size of ethylene plants has increased rapidly until today an economic unit is of the order of 450,000 tons output per annum. This means that gasification or liquefaction plants capable of producing hydrocarbon are highly suitable for the production of ethylene. A very large gasification unit therefore will be required to match a modern ethylene plant, in respect of which plant manufacturers would do well to consider.

The by-products of coal conversion plants are also worthy of examination to consider whether they will affect the economics of the production of gas or liquids from coal. As has already been discussed, environmental control will force sulphur removal from either the stack gases or at an earlier stage during combustion, or conversion. There should therefore, be a theoretical market for sulphur, except for the fact that increasing quantities of sulphur will become available. We need look no further than the problems of the Polish sulphur mines some ten years ago, when strict control over sulphur emission to the atmosphere resulted in such a sulphur surplus that prices tumbled. There is considerable truth in the saying that the solving of one man's pollution problem's is often the finish of another man's business. There appears to be little joy ahead in sulphur from coal, unless sold at prices which make mined sulphur uneconomic.

The possible use in the production of Ammonium sulphate with its low nitrogen content and contribution to acidity of the soil does not offer good prospects for sulphur either.

Competitive Position of Coal

At first sight it might appear that coal will take over from oil - tomorrow, but a comparison between coal and oil may make the position clearer.

Coal. The impact of its natural constituents are not insignificant. Oxygen at about 10% by weight increases the amount of coal needed to be handled to produce hydrocarbons. It is accompanied by a fall in calorific value of the coal and in turn its ability to support endothermic reactions.

Sulphur removes hydrogen, as hydrogen sulphide, theoretically taking two atoms of hydrogen for every atom of sulphur, but in terms of weight the ratio is 1:16.

Nitrogen occurring naturally in the coal will form ammonia, if gasification takes place at low temperatures, accompanied by high pressure. Ammonia contains 20% hydrogen. On the other hand, high temperatures yield elemental nitrogen instead of ammonia.

Oil. By comparison oil contains around 1% less oxygen, despite a considerable variation in composition. It bears the same range of sulphur contents and the hydrogen to carbon ratio is about twice that of bituminous coal. Ash is virtually absent. While exploration is much more expensive, recovery of oil and subsequent storage are much simpler. This offers considerable advantage over coal, but the price also needs to be considered.

Natural Gas. When hydrogen is needed in abundance, as in the case of ammonia, methanol and other chemicals, methane scores heavily. If the range is extended to include ethane and heavier fractions they too provide an ideal raw material. Always available in the States, the North Sea fields such as Brent now offer an abundant supply. Plants needed to take this type of feedstock also cost less to construct.

Coal, however, does not compete well in the production of ethylene simply because the route from synthesis gas does not exist, in commercial terms. On the other hand it is possible to reach some of the end products of ethylene, starting from acetylene. Between now and the time when coal takes on a major role in the provision of feed for the chemical industry, a lot may be done through research and development, to bridge the gap between the two.

It is also reasonable to assume that with a considerable amount of research taking place throughout all the developed countries in the world, aimed at conserving energy use and finding substitute sources through nuclear energy, as well as renewable sources, the rate of growth in the use of chemicals which are hydrocarbon dependent will increase much faster than the rate of energy increase. This however overlooks the very considerable effort being put into recycling of plastics, the recovery of what were formerly waste materials or changes in technology which will reduce their need. Trichlorethylene is a good example in addition to many other solvents, which in the days before the oil crisis and pollution control, were dumped. This rarely happens today. Ethylene-glycol is one exception, simply because it is much easier to open the cock and drain the radiator than take steps to recover the anti-freeze. Technological advances will add further to these pressures, although the level of world unemployment may well change this trend as steps are taken to keep people at work who might otherwise become redundant. Earlier a situation in the U.S.S.R. was described in which the coal from the hydraulically operated mine was transported to a complex where preparation and chemical production units shared the same site, well away from the mine. Similar planning is proposed for Selby, the new British mine. This arrangement involves more coal being processed than would otherwise be the case for one plant by itself. If a generating station were included, there could be problems arising from the fact that production at the chemical complex would need to be continous whereas the power station would follow a pattern of demand. Integration would certainly present difficulties.

Coal Tar Production

Coal Tar stems traditionally from three sources:

- production of metallurgical coke
- carbonisation of coal
- fixed-bed gasification of coal

Gas effluent is often dirty, giving rise to a risk of fouling of the heat exchanger; it is therefore quenched with water sprays. The gas is then cooled further with more tar, oil and water condensing out. The following group of by-products may be recovered.

Ammonia. As has been pointed out the nitrogen content of the coal is often recovered in the form of ammonium sulphate. Equally, it could be as anhydrous ammonia or ammonium phosphate.

Examination of methods used in a number of countries has highlighted three processes for the recovery of ammonia.

1) Gas is passed through sulphuric acid at a temperature above the dewpoint of water. Because the method results in considerable amounts of corrosion involving loss of quality of both the tar and sulphuric acid, it is rarely used.

2) The ammonia in this case is absorbed in aqueous scrubbers and stripped, after which it is absorbed in the sulphuric acid or else left as ammonia in liquid form at a commercial concentration.

3) A method in which the gas is cooled down to about 30°C and then passed through an electrostatic precipitator is most used. Tar is removed in the final stage. After this, the gas is reheated above the dewpoint to about 60°C so that condensation does not take place in the saturator. This is where the gas is usually passed through the sulphuric acid. The ammonia also passes to the saturator, the gas carrying mainly weak acidic compounds going on for oil recovery and removal of the sulphur. Ammonium sulphate crystals are formed, their size depending upon the design of saturator.

Tar bases. These by-products consist of nitrogen carrying heterocyclic compounds such as pyridine. These constituents with a high boiling point become condensed with the tar and with the flushing liquors. The lower boiling point compounds which tend to be sought as chemicals, are removed from the gas either at the stage of ammonia removal or that of light oil. By-products are removed continously, neutralisation with ammonia being carried out. As a result separation occurs into crude pyridene and a layer of ammonium sulphate, which is recycled.

Tar acids. The ammoniacal liquors contain tar acids and other compounds of an acidic nature - tar acids being mixtures of phenolic compounds. While coke oven gases contain mainly phenol, other acids derived from different sources may be made up of larger amounts of cresols and xylenols - cresylic acids. Tar acids may be recovered from the liquor by solvent extraction using benzene, a phenol-free tar oil fraction or some other convenient solvent. The latter is removed by distillation.

Tar acids may also be recovered from mixed coal and petroleum feedstocks, using caustic to absorb them or by the addition of carbon dioxide to increase the acidity. Solvent extraction with naphtha and methanol is then carried out to remove impurities.

Acetylene

For many years this gas was produced from calcium carbide; hydrolysis taking it through to acetylene. Coal carbonised to coke with lime, using an electric furnace provides the calcium carbide. The method is energy intensive and therefore not to be encouraged when the objective is conservation. New methods are being developed using coal. This would involve the injection of pulverised coal into an electric arc, in a hydrogen atmosphere, where volatiles form acetylene.

More recently British Petroleum through BP Chemicals have created a plant in Britain at Humberside for the production of acetylene, based upon the partial oxidation of methane. Naphtha can also be used, but there is no reason why coal derived liquids and gases could not be used too.

Conclusions

The Chemical Industry is a vital part of any industrialised nation. Without it, life as we know it today would grind to a halt. For the time being oil and natural gas including gas liquids will supply all the feedstock required. In fact, in terms of ethylene production, potential supply is ahead of demand. While the British Government is advocating four plants, the industry sees two only at the most, over the next few years, as being necessary. But that will change as demand exceeds the rate at which hydrocarbon can be extracted as oil or gas. Parallel with this, the conversion of coal to provide a feedstock is vital. The gas route is already here through the production and use of medium Btu gas with development for widespread use within ten years, while methane is already employed as a feedstock.

The production of liquids from coal will take longer. Their use it is unlikely on a commercial scale this side of 2000.

Conversion is also a step forward so far as transport is concerned. Despite considerable advances in movement by hydraulic means gas offers particular advantages except in the low Btu form. Pressure must be maintained through the use of compressors but unlike liquids no pumping is required. Both can be stored more easily than coal, gas being capable of liquefaction before being pumped into underground caverns. This is particularly applicable in countries possessing salt mines, use already being made of this method in a number of European countries, particularly France and Germany.

The Role of Coal in tomorrows technology is vital. Coal will clearly be indispensable.

Without doubt the British Government takes this view, having at the end of May 1978 announced a £43 million programme to achieve the substitution of coal-derived fuels for natural oil and gas by the end of the century. The programme is to run for eight years, £20 million being invested by Government over a period of five years.

Three projects which will begin immediately are:

- research into liquid solvent extraction or the extraction of petrol from coal at a cost of £1 billion.
- supercritical gas solvent extraction or the production of chemical feed from coal costing £15 million.
- fixed bed composite gasification or the production of substitute natural gas from coal at a cost of £12 million.

These three projects are all extensions of work already described at NCB's Stoke Orchard Research Station and the British Gas Westfield project.

These proposals stem from a report by the coal industry's tripartite working party on research and development. The report suggested that the figure of 170 million tons of coal, the production target for the end of the century, might well need to be raised considerably, since 80 million tons of coal would be needed to produce sufficient petrol substitute to meet ever present petrol consumption.

Longer term projects include fluidised bed combustion for the production of electricity, the prototype of the technology having been researched by Babcock and Wilcox.

The objective is clear: the technology must be developed as quickly as possible now that the money is available.

Chapter 11

ENVIRONMENTAL CONTROL

The mining of coal inevitably brings considerable risk with it from natural forces which in other forms face the deep sea fisherman, the steeplejack and those dealing with infectious diseases. But the miner is underground: unless very considerable precautions are taken he will become entombed, due to subsidence of the soil, gassed or asphyxiated due to sulphur fumes, lack of oxygen and the presence of methane. Blown up if there is excessive dust or a combination of oxygen and methane occurs or drowned if there is too much water.

Clearly rigid control is necessary to ensure that these situations do not occur. Steps must be taken to monitor regularly the presence of gases; that air is permitted to circulate; that water is removed quickly or else put to work cutting and moving the coal, or reducing the working temperature at the coal face. The presence of dust - the greatest potential killer of all even in the absence of explosions - due to the effect on miners' lungs - must be monitored and controlled.

While scientific progress has enabled most of the potential hazards to be reduced considerably, ironically advances in engineering and mechanical recovery techniques have resulted in a considerable increase in dust. As can be seen from chapter 4, a great deal of research is underway through the use of lower speed shearers and design of cutters, water and dust filters, to overcome the problem without losing production.

But once the coal has been dug, the environmental problems move to the surface. The making safe of existing slag heaps, the distribution of dirt so that new heaps are not created, is of paramount importance. The working of mines to leave pillars at regular intervals as at the new British mine at Selby, where a river below sea level and historic buildings such as the Abbey might otherwise be damaged were subsidence to occur, was a precondition of planning permission.

The siting of preparation plants away from the pithead is also being insisted upon by the planners, often to good advantage so that the clutter of vehicles converging upon one point is removed. This new thinking, already seen in chapter 3, where coal is removed to a complex which includes chemical manufacture, is made all the more practical by the use of drift mines, that is the use of a tunnel instead of a shaft for removal of coal. New possibilities have been opened up through the use of hydraulic transport, also discussed earlier.

Even then, the effects of coal upon the community, far removed from the pits, must be controlled.

This may be achieved by the type of coal which is permitted to be burnt for domestic purposes in built up areas and so limit the emission of sulphur to the atmosphere. It is this strict control in London which has done so much to eliminate the infamous pea soup fogs so common during winter there during the first half of this century.

For industry, the problem is slightly different. Clearly sulphur free coals would be too expensive in many parts of the world to use on their own. As has been seen in chapter 2, the East Coast of the United States enjoys plentiful resources, the bulk of which are of high sulphur content. The use of scrubbers to reduce emissions amounts to about 6% of the cost of the plant, in the case of gasifiers and therefore the use of fluidised beds with the addition of lime, which will bring the sulphur out with the ash, is being developed.

The design of grate also is crucial to the efficient burning of a wide range of coals so that the equipment is able to withstand the potential corrosion properties of the coals.

HEALTH & SAFETY UNDERGROUND

One of the most developed degrees of Trade Union involvement in Health and Safety matters is to be found in British mines, where worker involvement has seen a dramatically changed situation.

It is clear in retrospect that the majority of accidents do not come from what are commonly called disasters. They result from the effects over a period of working in the sort of environment described at the beginning of this chapter. That is an environment which is clearly unnatural and where the circulation of air is needed to dilute harmful and flammable gases and airborne dust and where artificial light and the hazards which go with producing electricity in an atmosphere in which there is a risk of fire. It is this aspect rather than the absence of daylight which constitutes the unnatural unenvironment.

Other problems are to be found such as varying temperatures and humidity, fire and explosions and the need for support for workings and roadways. One of the greatest problems of increased mechanisation apart from dust is noise. Until recently immense physical effort was needed to move supports, but even now supports must be taken up to the coal face and assembled by hand.

The British experience is the easiest to recount because it is to hand. It is also a good example of what can be achieved by co-ordinated effort amongst groups of workers.

Legislation

Between 1908 and 1912 in the United Kingdom an average of 5,729 accidents involving injury to people occurred annually. However, in 1906 legislation had been set up which required the reporting of accidents. And so it was heartening to find that by 1932, the annual average for the years 1923 to 1932 had fallen to 4,197 injuries. Too many, but moving in the right direction.

However, since 1932 there has been a steady decrease in the number of accidents, as can be seen from Table 39 where the number of people killed or seriously injured in coal mines is recorded.

TABLE 39 Numbers Killed or Injured 1933-76

Year	Killed	Seriously Injured	Rate per 100,000 manshifts
1933-42	877	3,123	
1943-52	538	2,293	
1958	327	1,752	
1965	216	1,159	1.31
1966	160	1,061	1.26
1967	151	982	1.25
1968	115	851	1.22
1969	100	712	1.16
1970	91	641	1.14
1971	72	641	1.11
1972	64	519	1.10
1973	80	553	1.14
1974	48	497	1.12
1975	64	586	1.20
1976	49	535	1.11

The rate per 100,000 manshifts column in Table 39 is perhaps a better guide to progress than those injured, since this takes into consideration the fall off in numbers employed in the mines. For instance, in terms of numbers killed or injured 1971 appears to be a poorer year than 1976 - that is, more accidents - and yet in terms of numbers employed it was similar to 1976, although the total was much lower.

However, returning to 1906, the Notices of Accidents Act of that year was extended to include a number of events whether or not personal injury had occurred:

- all cases of ignition of gas or dust below ground, other than ignition of gas in a safety lamp.
- all cases of fire below ground.
- all cases of breakage of ropes, chains or other gear by which men are lowered or raised.
- all cases of overwind of cages while men are being lowered or raised.
- all cases of inrush of water from old workings.

The Coal Mines Act of 1911 and its accompanying regulations provided the first comprehensive legislation governing the health and safety aspect of coal mining.

However, during the interwar years several of the miners' unions together with the Coal Owners' Associations, a system of inspections was set up as a result of which the mines were inspected on behalf of the workforce by Safety Board Inspectors. The cost of such inspections carried out by the Safety Board Inspector, together with a local mineworker, was borne by the coal owners and Trade Union jointly.

The first mining engineer was appointed in South Wales just before the war, but by 1952 it would appear that only the sixth professionally qualified man had been taken on. In 1957 the Mines & Quarries Act of 1954 had become implemented. Amongst the provisions was one for the majority group representative of the total number of people employed at the mine to nominate somebody to accept notices on behalf of employees.

It also became a requirement that where death or serious injury resulted from an accident, certain information had to be provided:

- Fracture to the arm, forearm, thigh or leg, pelvis, skull and spine
- Dislocation of elbow, hip, knee, spine or shoulder
- Amputation of a foot or hand or part of either
- Loss of sight of an eye
- Other serious bodily injury such as asphyxia, burns, internal haemorrhage or other injury likely to endanger life, cause permanent injury or disablement

The Notification of Dangerous Occurrences Order 1959 specified those incidents which had to be reported.

a) Any gas, other than gas in a safety lamp or dust is ignited below ground.

b) Any gas in part of a firedamp drainage system on the surface or in any exhauster house is ignited accidently.

c) Any fire breaks out below ground.

d) Any person, in consequence of any smoke or other indication that a fire may have broken out below ground, has been caused to leave any place pursuant to paragraph (1) of Regulation 11 of the Coal and Other Mines (Fire and Rescue) Regulations 1956 or section 79 of the Act.

e) Any fire breaks out on the surface endangering the operation of any winding or haulage apparatus installed at a shaft or unwalkable outlet, or any mechanically operated apparatus for producing ventilation below ground in the mine.

f) There is a violent outburst of gas together with coal or other solid matter into the mine workings except when such an outburst is caused intentionally.

g) There is any inrush of noxious or flammable gas from old workings.

h) There is any inrush of water or material which flows when wet, from any source.

i) Any rope, chain or coupling or other gear of a similar kind by which persons are carried through any shaft, staple-pit or unwalkable outlet breaks.

j) Any rope, chain or coupling or other gear of a similar kind used for the transport of persons below ground, breaks.

k) Any cage being used for the carriage of persons is overwound or any cage not being so used is overwound and becomes detached from its winding rope; or any cage operated by means of friction of a rope on a winding sheave is brought to rest by the apparatus provided in the headframe of the shaft or part of the shaft below the lowest landing for the time being in use, being apparatus provided for bringing the cage to rest in the event of its being overwound.

l) There is any explosion from or collapse or burst of any apparatus which is used for generating a pressure of air, gas or steam greater than atmospheric or for storing air, gas or steam at such a pressure.

Provided that the provisions of this paragraph shall not apply to any explosion from any boiler to which the Boiler Explosions Act 1882 applies.

m) Any ventilation apparatus other than an auxiliary fan breaks down causing a substantial reduction in the ventilation of the mine lasting for a period exceeding thirty minutes.

n) Any headframe, winding engine house, screen or tippler house, or vehicle gantry collapses.

o) Any mechanically powered crane, grab or other lifting machine overturns, or any part thereof (other than the hoisting rope of a stock pile gantry crane) fails and causes the load of the machine to fall or, in the case of such a machine with a jib, the jib to fall.

p) Breathing apparatus or a smoke helmet or other apparatus serving the same purpose, while being used, fails to function safely or develops a defect likely to affect its safe working.

q) Forthwith after using and arising out of the use of breathing apparatus or a smoke helmet or other apparatus serving the same purpose, any person receives at the mine first aid or medical treatment by reason of his unfitness or suspected unfitness.

r) Any person suffers such an electric shock or burn, by coming into contact with any conductor, in a circuit in which the voltage for the time being exceeds twenty-five, that he receives therefore at the mine first aid or medical treatment.

s) Any person suffers such bodily injury, resulting from an explosion or discharge or any blasting material or device as defined in sub section (4) of section sixtynine of the Act, that he receives therefore at the mine first aid or medical treatment.

t) Any movement of material or any fire or any other event indicating that a tip to which Part 1 of the Mines and Quarries (Tips) Act 1969 applies is, or is likely to become insecure.

u) Any apparatus is used (other than for the purpose of training and practice) which has been provided at the mine in accordance with any scheme made pursuant to the mines (Emergency Egress) Regulations 1973 or any other arrangements are carried out in accordance with the scheme whereby persons employed below ground in the mine use means of egress therefrom in an emergency.

In addition it became an offence for anybody to disturb the place of the incident before three days had elapsed or in the event of a visit by inspectors, if that occurred first - that is H.M. Inspectorate and a Workers Inspector. Certain qualifications were made later to cover accidents to vehicles, etc. which otherwise might block the mine and itself become a hazard. But the terms had to be very strictly complied with before any movement of anything on the site was permitted. In the case of fire, the Fire and Rescue Regulations require that when a fire or the sign of a fire occurs workmen below ground must be withdrawn from all points likely to be affected by the fire, following which the area must be inspected by the manager and the appropriate Inspector.

The National Union of Mineworkers

This organisation was formed in 1945, since when the National Executive Committee has created a sub-committee to deal solely with matters of safety and health. This committee discusses trends and developments in these fields and advises the National Executive accordingly.

A National Safety & Engineering Department was set up in 1959, headed by a mining engineer, with an assistant appointed in 1976.

Eleven mining engineers and three electrical engineers, two of whom are also qualified mechanical engineers, are now employed by the "Areas".

Potential Hazards

Recognising the need to reduce risk the Union has given its support to:

- implementation and maintenance of support systems
- implementation and maintenance of safe transport and haulage systems
- provision of necessary tools and appliances
- prevention of misuse of equipment
- ensure proper training
- monitor environmental problems

The package and interpretation of legislation affecting the mines is such that the requirements of the 1974 Health and Safety at Work Act are not considered to meet the special needs of mineworkers. Coal mines have therefore been excluded from the provisions of the Act.

It should be made clear that this chapter has been discussing mineworking and protection of mineworkers as opposed to other activities of the National Coal Board.

Clearly the British legislation is aimed at both Health and Safety, although emphasis has been placed rather more on the Safety aspect than Health. The National Union of Mineworkers is extremely concerned about both aspects. However, to obtain a view which places a greater emphasis on Health it may be helpful to look at the French Industry.

A FRENCH VIEW

The post second world war era showed the cumulative results of poverty, poor working conditions and ignorance of the most elementary rules of industrial hygiene. Tuberculosis, silicosis and physiological problems were rife according to the Chief Medical Officer of the Charbonnages de France. The labour force was unable to cope with the exhausting work of digging for coal by hand.

Conditions have now changed to such a degree that the life span of the miner is about that of the rest of the population. Ankylostomiosis and nystagmus are now almost unheard of, while silico-tuberculosis now occurs very infrequently and is less serious following the introduction of anti-biotics. The incidence of pneumoconiosis has fallen from between 1 to 3% of the workforce in some areas to between 0.2 and 0.3%. Further progress appears to be difficult, perhaps with the arrival of shearers and similar dust producing forms of mechanised miners. Other reasons may be:

- The precise cause of coal miners pneumoconiosis is not totally understood since it has proved impossible to develop mining conditions of zero dust.
- The main dust suppressor is still water, which when used to excess provides its own problems for those working in the area.

Although detection of new cases may be much smaller it must not be forgotten that within the Community there are probably some 15,000 ex-miners permanently at risk - and possibly more. Other complications such as chronic pulmonary heart and acute respiratory insufficiency show little response to treatment. It is said that there is no known cause or treatment for progressive massive fibrosis - PMF - which often makes worse miners pneumoconiosis, the main cause of other problems such as chronic pulmonary heart and acute respiratory impairment.

The real need is to perfect technical prevention methods and to do all possible even to the extent of banning forms of mechanisation producing dust which cannot be controlled.

Secondary Hazards

These by comparison with the respiratory problems are less difficult to monitor and control. They include:

a) Noise which should not be judged by normal levels of measurable decibals alone but also the time of exposure during the shift.

b) Vibration from manual tools which affect the hand and elbow or machines which may cause the whole body to vibrate.

c) Heat resulting from the depth of the seam and also as more powerful motors are introduced.

d) Strenuous physical exertion needed to move heavy equipment.

e) Working in unnatural positions causing backache, bursitis and meniscitis.

One aspect being investigated is the effect of shift working necessary to obtain a return on expensive machinery upon the health of those who must adapt their life cycles to this constantly changing background. Here of course the miner is no different to the industrial shift worker, although the latter normally works in far better conditions.

The future will present new problems, mainly :

- Toxicological aspects as new chemical substances, particularly organic compounds - polychlorinated bi-phenols, polyurethene and resins, urea formaldehyde foam or isocyanates and many others, find their way underground as new plant and equipment are introduced.

- Monotony, which may lead to carelessness.

- Brain fatigue which could also result in carelessness.

- The incidence of other effects of dust and mining conditions, such as carcinogesis which formerly was prevented from developing since pneumoconosis was dominant.

REDUCTION IN RISK

One way of reducing risk from dust is to remove men as far as possible from the scene of coal fracture and therefore coal dust.

This may be achieved in three ways:

1) Remote control techniques.

2) Extraction methods.

3) Heavy duty equipment.

Remote Control Techniques

These may be applied to mechanical methods as well as hydrological means of extraction, both of which have been described in earlier chapters.

Extraction Methods

These provide good examples from British experience where the web depth has been increased from a maximum of 22in to one metre width. Output per shear has increased by 65%, while output per man shift has gone up by 40%. The rate of compensable accidents per 100,000 manshifts fell from 300 to 70 and the average levels of respirable dust (milligrammes per cubic metre) were reduced from 13.3 to 8.8. Of equal importance, the shearer cutting distance was reduced from 123 km to 75 km for every 100,000 tons of coal produced.

Heavy Duty Equipment

This category includes some of the equipment already discussed. Wider web extraction practice extended to greater heights will require supports having a capability of about 400 tonnes to obtain the same resistance level to movement of the strata as is the case with conventional faces.

Considerable importance is attached to the elimination of power loader haulage chains. These have been a source of accident in the past, as well as proving unreliable.

Dust Control

This is being achieved by the use of slower rotating drums on shearer loaders mounted with larger but fewer cutters. Progress here may be judged from Table 40.

TABLE 40 Work Places Meeting Dust Standards March 1977

Working Place	No.	Percentage meeting Standard
Longwall Coal Faces Production Shift	759	99.2
Preparation Shift	82	100.0
Cutting Shift	25	100.0
Bord and Pillar Operations	55	96.4
Drivages	773	98.5
Intake roadways to coal faces	759	99.7
Transfer points In return airways	155	100.0
Loading points In return airways	14	100.0

Table 40 shows a reasonably satisfactory situation, best understood when it is realised that nearly three quarters of coal faces now have dust concentrations below 5 mg per cubic metre.

Clearly new equipment is merely changing the nature of the problem from one of less physical effort to the possibility of increased hazard from dust. Research is playing its part in reducing these problems.

HYDRAULIC MINING

The winning of coal as has already been described is by means of water monitors remotely controlled from a safe point in the gate road.

The fact that no access is required to the face effectively removes possible sources of accident, there. There are a number of other factors which contribute to increased safety.

Hydro-Winning and Hydro-Transport

These factors can be seen to include:

Coal Face

Absence of men and machinery removes the possibilities of:

falls of coal and rock - machines - conveyors - tools - stumbling, falling or slipping - falling or sliding objects.

Gateroads

A reduction in hazards from strata pressure and stresses by the adoption of retreating systems and sub-level caving. As a result reduced support loading is needed and there is a reduced likelihood of roadway failure.

Also reduced hazards from machinery and transport, the latter being by flume and possibly pipe. No electric drives or other apparatus are required, so eliminating this type of hazard.

Outbye Workings

The absence of mineral transport other than by continuous pipe systems reduces this type of danger, as does reduced need for the movement of materials.

Fire Hazards

This usual danger is considerably reduced due to:

- use of water for coal winning
- separation of coal and rock during recovery
- retreat mining methods
- lower level of coal dust deposits both from hydraulic transport in flumes and pipes

Hydraulic Transport of Stowing Dirt

There are a number of factors which result in increased safety associated with this activity, which may be divided into four fields:

Transport from Surface to Production Area

The use of a continuous pipe system reduces the normal hazards of this operation.

Face

The danger of rock falls resulting from stowing and therefore improved roof conditions, is reduced, as well as that normally associated with the flushing of material from the caved goaf. Greater use of stowing also lessens the risk from firedamp emission from the goaf and therefore the likelihood of explosion. This results in improved efficiency of firedamp drainage. The greater use of stowing also contributes to reduced fire hazard.

Gateroads

Other factors here which contribute to safety also include roadway stability, again as a result of stowing and gateside packs. Also as seen earlier improved efficiency of firedamp drainage reduces the likelihood of explosion.

Maintenance

Greater use of stowing and gateside packs also results in the need for less roadway repair work.

Subsidence Damage

This is greatly reduced as a result of increased use of stowing material.

Benefits to Health

Three other features beneficial to health result directly from hydraulic mining. Cooling or reduced heating effects result from the use of water for coal winning and transport at high ambient temperatures. This applies particularly to the ventilation air. The greater use of stowage also results in a fall in dry-bulb temperature in hot mines of around 3°C.

Similarly the virtual absence of machinery and moving equipment, apart from pumps, produces a considerable abatement in noise nuisance.

The lack of need for men to work at the face in a kneeling position is a further benefit to health of workers underground.

NCB MEDICAL SERVICE

The National Coal Board's medical service reports other aspects of health and safety underground and the effect of working there.

It was recently agreed that the retiring age for men with more than twenty years working underground would be reduced gradually over the next few years. This must add considerably to the number of routine medical examinations needed as men leave under this scheme and others come to replace them. Bearing in mind the extent of wastage experienced with recruits the burden will exceed considerably the numbers indicated in Table 41 which shows the number of men in each age group involved, as at March 1976.

TABLE 41 Number of Men aged between 60 & 64 as at March 1976

Age	Numbers Employed - thousands
60	4.9
61	4.9
62	4.7
63	3.7
64	2.9

Table 41 shows the number of men aged 60 and over as representing some 12% of the total numbers working underground. If the age range were to be increased to include all the over 50's, which is the case in the U.S.S.R., the numbers involved would rise to about 38% of the total.

Research Projects

Research in general is very largely the responsibility of the Institute of Occupational Medicine sited at Edinburgh. NCB Medical Service is at present examining three areas:

Hand Injuries - Exercise Tolerance Tests - Causes of Progressive Massive Fibrosis

Hand Injuries

Mining operations are of a nature causing a large number of injuries to fingers and hands - forming between 20 and 25% of all accidents leading to absence from work for three or more days. The main course is to study the way in which hand injuries are caused and how they may be prevented. This will include examination of the design of equipment and the possibility of producing protective gloves.

Exercise Tolerance Tests

An improved test is being developed in part for use in the annual medical examination of rescue men.

Cause of Progressive Massive Fibrosis

Two research projects are nearing completion, one involving the cause of progressive massive fibrosis - particularly the role of tuberculosis - while the other involves the x-ray appearances of pneumoniosis in different coal fields mining different types of coal.

Underground Coal Gasification

The obvious attractions which UCG offers environmentally constitutes the contribution which this method can make to the Quality of Life in bringing men up to the surface from underground. This was the feature that attracted Lennin in the early 1900's. However UCG is not all bonus since transfer of work from the mines to the surface disfigures the landscape.

The degree to which in-situ is acceptable environmentally depends upon the area chosen and local concern, which will vary from one country to another. In the Wyoming prairie few complaints will result but the effect upon water supplies will be quite different.

It is therefore difficult to review the environmental aspect in a world context. A European scene is probably easier.

Siting

Transport costs of low Btu gas demands that the system using the gas shall be near to the "in-situ" site. Power stations are not welcomed, which may create problems.

Equipment to be considered when choosing a site includes:

- drilling rigs and associated equipment
- compressor buildings
- pipe lines
- flare stacks
- workshop and stores
- control and administration
- welfare departments

Those which invade the skyline are unlikely to be accepted unopposed.

At least three drilling rigs will be operating on the average site moving across the area as new boreholes are opened up.

Excess substandard gas will need to be flared. While the flame and plume will be clean for most of the time special burners may be necessary to ensure complete combustion of smoke during start up and certain other periods, conducive to the production of smoke. An enclosed incinerator may be required anyway to allay local objections.

Surface pipework covering several hectares will be unavoidable and therefore design must play a particularly important part in a successful layout. Good drainage and access for the rigs is vital.

However, none of the normal disfigurement associated with conventional mining such as coal heaps or transport associated with its movement will be present. The problems to be met will depend upon the method used:

<u>Vertical drilling</u>. This will involve movement of rigs over the whole site placed at around 100 foot (30m) intervals covering some half of the site. The top soil may therefore require ultimate cultivation achieved best either by its temporary removal to prevent contamination or the introduction of sewage sludge at a later date when the whole area is restored.

<u>Preliminary mining</u>. In this case the spaces between the holes would be similar to the vertical method, but in addition the rows are to be found about 900 feet (275m) apart. The areas between the rows should still be available for limited agricultural purposes.

Over the period of the life of the site, considerable areas of land will be needed. That applies to a 100 MW power station, but were a 1,000 MW station to be involved the problems from all aspects would be greater.

<u>Health & safety</u>. There are other factors to be considered which in countries such as Britain are controlled by legislation such as the Control of Pollution or Health and Safety at Work Acts. In the case of the latter, everybody has a

responsibility towards third parties, including the public. These potential sources of problems can easily be identified in advance since they apply to many other aspects of industrial activity. They include:

Dirt. UCG produces very little emission. The ash from combustion is mostly left in the seam area. On the other hand subsoils produced from the cores estimated to reach 70,000 tons over 20 years - on a 100 MW site - must be disposed of, but this amounts to only around 10 tons per day.

Chippings must be screened out. Some of the material will be returned underground.

Fire. Experience shows that the danger of fire is remote. It is rather more a question of keeping the combustion front going. Any risk of fire can be controlled by cutting off the oxygen supply in what amounts to a sealed system, by comparison with the normal warren of passages in a conventional mine.

Special precautions can also be taken to surround the gasifier with boreholes for introduction of fire quenching fluids or as a last resort, the whole gasifier could be flooded.

There is also the possibility of reducing the residual heat and therefore any likely risk, by recovery, for some low temperature heat use.

Gas emissions. The gases to be emitted are bound to be of a highly toxic nature due to the presence of carbon monoxide, which is a cumulative poison and which even at very low concentrations can be dangerous with long exposure. This is unlikely to present a problem when the site is open to the weather. There is little theoretical risk of leakage to the surface from levels in excess of 100 feet - 30 metres - although fissures could make this into a hazard.

Smell. During the process of carbonisation of the coal some tars and phenolic by-products will be produced. High sulphur coals, particularly Eastern coals in the United States will produce sulphur dioxide as well as Hydrogen Sulphide, resulting in offensive smells. The gases would need to be scrubbed, possibly with some payback from the sulphur recovered.

Surface heating. Experience has shown that due to the low thermal conductivity of the rock, heat escapes to the surface at a very slow rate, the temperatures noted being negligible at distances of more than 50 feet (15m) from the reaction zone.

British experience in opening up the Newman Spinney site some four years after the finish of the tests stopped that, the reaction zone was still hot, but the heat was contained very close to the original gasifier. The conclusion reached was that at depths greater than 300 feet (90m) little change in temperature is likely to be experienced at ground level. More information should be available from the Laramie Energy Research Centre test site at Hanna, Wyoming.

Subsidence. There is clearly a difference between UCG and conventional mining, even though coal is removed in both instances. With UCG the rocks adjacent to the seam will have been affected by the heat reaction, which could result in a different pattern of subsidence. Whether the effect of heat will strengthen some rocks or weaken others must be in doubt until more evidence is available.

Water pollution. There must always be a danger of water pollution if the reaction products leak out of the area of gasification and into the catchment area. Conditions at Hanna approximate closely to this situation, where the seam being gasified is an aquifer and water is scarce.

This situation demands regular monitoring. The main problem where water is used for drinking and chlorine added as occurs in many countries is that even minute traces of phenol create a taste, which is totally repulsive.

In general terms few hazards are normally associated with UGC, provided the oxygen is handled with care.

The accident rate for Vertical Drilling is similar to experience with oil drilling on land. Preliminary Mining will involve a limited number of men underground working on development of the roadway and employing the usual type of tunnelling machines. Normal mining standards referred to earlier in this chapter must be employed in this situation.

Manned Control Underground

There may be instances in which it is necessary to have a manned control base below the surface. Although not near the extraction area they could be faced with high temperatures and chemical contamination from the particular underground process. Local situations, such as refrigeration and dust suppression may be the most economic solution. Recycling the air is a possibility. Here, dust, heat and moisture are removed, although Firedamp remains. It is suggested by B.P., Esso and the Microbiological Research Establishment that this may be dealt with successfully by catalytic or microbiological oxidation. In poor conditions complete air conditioning may be practical and economic.

There will always be some need with Preliminary Mining to move along the roadways where Firedamp is likely to be the main problem. A number of suggestions have been put forward to deal with this problem including:

- microbiological digestion of methane in the seam which has been investigated in U.S.S.R.
- drainage of the area using surface boreholes reinforced by hydrofracture
- operating the whole mine at 2 bar pressure, a method examined in the United States, not really applicable to a conventional mine

Mine Emergencies

Safety in coal mining operations will be enhanced considerably as a result of a National Coal Board decision to add to its existing emergency equipment a new mine escape unit designed by AGA Spiro, in conjunction with its Swedish associate.

Most escape apparatus has proved unsuitable in situations where there is a deficiency of oxygen and where men must run fast to escape from an abnoxious atmosphere.

Any equipment to be introduced must be light and simple in design so that it could be used with minimum training. It also has to be self contained and capable of providing sufficient oxygen at all times at prevailing levels of air temperature. The escape set, with the oxygen cylinder fully charged weighs only 6.5 Kg. It has a protective case assembly of anti-static glass fibre reinforced plastic, the overall dimensions being 430 mm long, 360 mm wide and 110 mm deep. Breathing is through a mouth-piece carrying inhalation and exhalation non-return valves. Exhaled gas passes through a soda line filter media into the inhalation bag and delivered cool to the mouth-piece. The constant flow of oxygen is directed into the demand valve connected to the exhalation bag where demand supply is also available. The system has been approved by both the NCB and Inspectorate, meeting the requirements of the British Health and Safety Executive.

The objective in this chapter has been to examine the problems which exist and which affect both Health and Safety and the ways of reducing risk.

In part, this can be met by introducing legislation, but for any law to be effective the cause of the problems must be understood and the ways in which men may be protected from potential hazards. This is why it was felt to be important to review the medical scene and the work being carried out there.

The review of in-situ gasification should indicate the attention which modern mining practices must pay to the environment so that it is not spoiled by unsightly landscapes, but rather that operations should merge. When opencast does take place, it is vital that the land becomes restored afterwards providing if possible a sight more pleasant to the eye than before mining began.

The goal is clear: there is still a long way to go.

APPENDICES

There are nine separate sections to this appendix ranging from A to I inclusive.

These sections are included to provide additional information which should enable the reader of this study to obtain a more complete picture of the role which coal already plays to-day. Also, the organisations available to ensure that the latest technology will be available in the years which lie ahead to provide the best environment and equipment possible for those who work in the industry.

A. Energy balances in coal producing countries

B. U.K. Mining Research and Development Establishment projects

C. European Community - priorities in remote control and monitoring R & D projects

D. Mines Safety and Health recommendations (13th Report, 1975) - European Commission

E. Organisations with mining-orientated activities

F. Petroleum Feedstocks and some of their derivatives

G. Sources of Information - coal producing countries

H. German Industrial Involvement in Coal Research and Development

I. Major suppliers of mining equipment

APPENDIX A

ENERGY BALANCES IN COAL PRODUCING COUNTRIES

The following energy balances expressed in millions of tons of coal equivalent have been extract from the 1978 OECD energy balances and are in respect of 1976. It should be noted that because electricity is brought in to some countries to supplement production from a variety of fuels, an asterisk is used instead of a percentage. A figure comparable to the other percentage imports would be meaningless.

Energy Source	Production	Imports	% Imports
Canada			
Solid Fuel	22.21	15.90	41.72
Oil	121.23	57.71	32.25
Petrol	-	3.11	100.00
Gas	89.14	0.15	0.18
Nuclear	7.22	-	-
Hydro-Geothermal	88.38	-	-
Electricity	-	0.49	*
Australia			
Solid Fuel	71.93	-	-
Oil	32.47	13.17	28.88
Petrol	-	4.89	100.00
Gas	6.72	-	-
Nuclear	-	-	-
Hydro-Geothermal	6.81	-	-
Electricity	-	-	*
United States			
Solid Fuel	611.59	2.38	0.39
Oil	715.41	469.32	39.61
Petrol	-	119.45	100.00
Gas	713.51	36.01	4.80
Nuclear	72.89	-	-
Hydro-Geothermal	105.44	-	-
Electricity	-	1.48	*

Energy Source	Production	Imports	% Imports
Japan			
Solid Fuel	22.43	70.11	75.56
Oil	.97	389.16	99.99
Petrol	-	37.65	100.00
Gas	4.04	11.46	73.94
Nuclear	13.13	-	-
Hydro-Geothermal	34.05	-	-
Electricity	-	-	-
Spain			
Solid Fuel	11.64	5.58	32.40
Oil	2.78	76.32	96.48
Petrol	-	3.03	100.00
Gas	-	2.29	100.00
Nuclear	2.73	-	-
Hydro-Geothermal	8.13	-	-
Electricity	-	0.22	*

The EEC energy balance is also included for comparison purposes although the solid fuel contribution has already been examined.

Energy Source	Production	Imports	% Imports
EEC			
Solid Fuel	292.89	73.80	20.12
Oil	35.36	828.78	95.90
Petrol	0.65	164.96	99.60
Gas	223.60	86.12	27.80
Nuclear	34.49	-	-
Hydro-Geothermal	39.79	-	-
Electricity	-	4.78	*

APPENDIX B

U.K. MINING RESEARCH AND DEVELOPMENT ESTABLISHMENT PROJECTS

List of Projects

Coalface

Stable elimination
Face end systems
Wide webs
Shearer drum loading characteristics
Automatic horizon control of coalwinning machines
Horizon sensing
Automatic steering of fixed-drum shearers
Automatic steering of ranging-drum shearers
Control engineering
Hardware development
Pick force sensor
Chainless haulage
Steep-seam brake
Brake and clutch unit
Management of AFC advance
Scraper conveyor action
Cable handling
Powered supports
Caliper shield supports
Coalface hydraulics
Megatonne advancing faces
Packing
Face machine gas emission and ventilation
Strata behaviour around coalfaces
Powered hand tools

Roadway Drivage

NCB/Dosco in-seam miner
Radio control of heading machines
Impact ripping
Basic rock cutting
Hard-rock cutting
Drilling
Roadway supports

Underground Transport

Rope-hauled transport
Braking system for locomotives and manriding cars
Track standards
Haulage ropes
Materials transfer points
Free-steered vehicles and roadway surfaces
Linear-motor-driven manrider
Locomotives
Transport of material
Minerals transfer points
Earth leakage trip for locomotive batteries
Diesel engines
Winding

Coal Preparation

Dry fines screening
Coal washing
Automatic controls
Ash monitoring
Moisture monitoring
Sulphur monitoring
Handling of magnetite
Coal handlability
Disposal of tailings
Centrifuges for small coal
Research

Comprehensive Monitoring

Computer systems technology
MINOS for environmental monitoring
MINOS for coal clearance
Other MINOS developments
Automation of bunkers
Environmental instruments
Transducers
Nucleonic monitors
Radio communications
Recorders for underground use
Intrinsic safety developments
Investigation of power line disturbances
Plant monitoring
Reliability studies

Basic Studies

Geophysical detection of faults
Strata properties
Roadway stability
Firedamp prediction and drainage
Climatic studies
Dust control: general
Dust control on the coalface
Dust control at face ends and in drivages
Dust collection
Dust capture by fine water droplets

Provision of clean air for underground workers
Dust sampling
Noise control
Mobile power supplies
Erosion and corrosion of materials in hydraulic seams
Lubrication and wear
Cutting tools
Fracture properties of mining steels
Corrosion protection
Calculation of stresses around a coalface
Numerical simulation of a cage buffer
Analysis of stresses in a slurry setting cone
Parameters for optimum shearer efficiency
Assessment of computer language APL
Humidity

Testing

Roof supports
Mechanical testing
Laboratory testing of underground machinery
Hydraulic testing
Electrical testing
Non-metallic materials
Non-destructive testing
Mining machinery: surface trials
Mining machinery: underground investigations

General

Long-range planning
Technology transfer and training
Consultancy
Instrumentation
Workshops and manufacturing control
Design and Drawing Office
Photography

APPENDIX C

EUROPEAN COMMUNITY - PRIORITIES IN REMOTE CONTROL AND MONITORING R & D PROJECTS

Medium-term research aid programme - 1975 to 1980

Operations Underground

Field	Group	Designation	Comments
II		Development work	
	1	Conventional methods of driving roadways	Techniques of drilling, shotfiring and removal of material
	2	*Fully mechanised driving of roadways*	Heading machines which take the full section. Selective heading machines Mechanisation of supports. Transport problems related to driving of roadways. Integration of the equipment and organisation of the work. *Automation of heading machines.*
	3	New methods for rock-cutting	
	4	Large hole drilling	
	5	Sinking by boring	
	6	Cutting machines for very short faces	
	7	Mechanised driving of rise-headings (cross cuts)	
	8	Technical problems of ventilation	Improvement of climate and measures against methane and dust.

Field	Group	Designation	Comments
	9	Technical problems arising from roof pressure and strata supports	
III		Methane studies, climatic problems, rock pressure and supports	
	1	Presence, movement and release of methane	
	2	Degassing of seams and drainage	Development of methods of methane drainage.
			Pre-infusion from a distance.
	3	Mine climate	
	4	Valorisation and utilisation of methane	Underground storage of methane.
			Recovering methane from closed pits and old workings.
III	5	*Ventilation:*	
		(a) Measuring appliances	
		(b) Techniques	Use of analogue models.
			Control of methane emission on high performance faces.
		(c) Prediction and calculation	
		(d) Auxiliary ventilation	
	6	Rock mechanics	Applied rock mechanics.
			Formation of breaks and cleavages.
	7	Rock pressure and supports:	Strata bolting
		(a) Stone drifts	
		(b) Gateroads	
		(c) Face	
		(d) Face-ends	
		(e) Other excavations	
	8	Roof and floor control in various geological and working conditions	

Field	Group	Designation	Comments
	9	Prevention of rockbursts and bumps	
	10	*Powered supports*	Adapting to various geological conditions and methods of winning. Remote control and automation.
	11	New support systems:	
		(a) Face	Shield supports, etc.
		(b) Roadways	
		(c) Face-ends	
	12	Special problems relating to production in:	
		(a) Thick seams	
		(b) Steep seams	
IV		Methods of working and techniques of coalgetting	
	1	Workability of coal	
	2	*Winning in fully mechanised places:*	
		(a) Cutting machines and ploughs	*Improving equipment for controlling machines*
		(b) Powered supports	Adapting to various winning conditions.
		(c) Face conveyors	Development and application for transport of equipment.
		(d) Stowing and caving	*Mechanisation and remote control.*
		(e) Face crushers	
	3	*Fully integrated mechanised systems of production*	Problems of face-ends. Elimination of idle time. Programming individual processes and face operations. *Remote control and automation in the face.*

Field	Group	Designation	Comments
			Development of high power machines requiring little maintenance and repairs.
	4	Special problems of working:	
		(a) Semi-steep seams	
		(b) Steep seams	
		(c) Thin seams	
		(d) Thick and very thick seams	
	5	New methods of winning	By hydromechanical methods.
			By auger mining in level seams.
			In steep seams.
			In thick seams.
	6	Technical problems of ventilation	Ventilation, improvement of climate and measures against methane and dust.
	7	Technical problems arising from rock pressure and strata supports	
V		Outbye services underground	
	1	Manriding	
	2	Transport of products and material	
	3	Mechanised loading points	
	4	*Automation of main haulage*	
	5	New transport systems	Monorails.
			Conveyor trains (Couloirs-roulants).
			Tyred vehicles, etc.
	6	Power supplies for underground operations	Higher voltage supplies.
			Cylindrical transformer, etc.
			Fire resistant liquids.
	7	Research and analysis on dynamic deformation in shafts	

Field	Group	Designation	Comments
VI		*Telecommunication, monitoring, remote control and automation*	
	1	*Telemetry*	
	2	*Monitoring systems*	*Perfection and development of mine control centres* *Monitoring equipment for underground electricity networks, etc.*
	3	*Remote control*	*Including interlocking systems for conveyors.*
	4	*Programmed control of individual methods of working and their equipment*	
	5	*Optimisation of integrated processes and operations*	*Ventilation.* *Reliability of electronic equipment.*
	6	*Automation*	*For example, main haulage systems.* *Ventilation.*

Operational Management and Planning

Field	Group	Designation	Comments
II		Planning, organisation and control of operations	
	1	Reduction in time of workings and improvement of the rate of utilisation of the machines	Elimination of idle time. Treatment of data received at the control centre.
	2	Improvement of ratio between nett and gróss output	
	3	Problems relating to the structure underground and at the surface	Simplification of the structure. Improvement of knowledge on interdependence of services. Improvement of cost system. Studies of underground and surface services.
	4	Assessment of coal reserves including brown coal	

Field	Group	Designation	Comments
	5	New planning techniques - integrated planning systems	
	6	Mathematical models	Selection of haulage methods.
	7	Optimisation of underground workings	Cross-section of roadways, lay-out of districts, face and roadway operations.
Preparation			
I		Mechanical coal preparation	
	1	Properties of raw coal and treated products	Evaluation of parameters measurable quantitatively.
	2	Stockpiling and homogenisation	
	3	Mechanism of the mechanical treatment process	Reduction of proportion of smalls and finest material by gentle crushing of grades sizes.
	4	Development of conventional techniques for coal preparation	Treatment of fines (flotation). Development of vacuum filtration
	5	Development and introduction of new methods	Electrostatic and pneumatic processes.
	6	*Control of the process*	*Control systems.* *Automatic rapid methods for determination of ash, moisture and sulphur content.* *Relation between parameters quantitatively measurable and parameters of uncontrolled techniques (i.e. not directly measurable).*
	7	Desulphurisation of the coal	
	8	Waste disposal	
II		Coking and briquetting of coal	
	1	Properties of coking coals and carbonisation products	*Automatic sampling and continuous preparation of samples* *Evaluation of parameters measurable quantitatively.* *Evaluation of qualities of coals e.g. size distribution of coke produced.*

Field	Group	Designation	Comments
			Widening of the coking coal basis.
	2	Mechanism of coking (pyrolysis)	Pyrolysis under extreme conditions.
			Quantitative research on behaviour of coal swelling in coke ovens.
	3	Development of conventional coking techniques:	
		(a) New methods of research	Determination of swelling pressure of coals.
			Thermal studies on regenerators of coke oven batteries.
		(b) Increased capacities and the productivity of coke oven batteries	Charging pre-heated coking coal.
			Blending charges.
			Refractory material allowing a reduction in the width of the stretchers (wall thickness), i.e. bricks.
			Increasing the temperature of the flues.
		(c) Mechanisation and automation of the processes	
		(d) Manufacture of foundry coke	
	4	*Thermal balance and process control*	Optimisation of oven heating.
			Reduction of heat required for coking.
			Influence of higher temperatures of walls (arches) on quality coke, swelling and productivity.
			Relation between the parameters and the variable technical parameters.
			Determination of temperature/time relationship.
	5	Production and purification of by-products and coking gas	Increasing the yield of simple aromatics from coal tar.

Field	Group	Designation	Comments
			Benefication of complex aromatics from coal tar for utilisation in the chemical and plastics industry.
	6	New methods of coking for coal	Continuous coking process of smalls and fines for the production of formed coke.
	7	New methods of briquetting brown coal	Based on rough coal, dried coal and small coke.
	8	Methods of manufacturing smokeless briquettes from coal	
	9	Technical and economic problems in environmental protection	Measures against emission during charging and discharging the ovens.
III		New chemical and physical processes and products from coal	
		(a) Fundamental research for new chemical and physical processes	
	1	Chemical constituion of coal	Studies of coal constituion and reactions involved in processing.
	2	Petrographic methods of analysis	Classification. Physical and chemical constitution; properties of petrographic constituents. Use in process control.
	3	Physical properties of coal	Physical properties as they affect the processing of coal.
	4	High intensity chemical reactions	High temperature and/or pressure reactions in pyrolysis, gasification, hydrogenation, etc.
		(b) New products and processes from coal	
	1	Gasification	Pressure gasification. High and low calorific value gas. Utilisation of heat from nuclear reactors.

Field	Group	Designation	Comments
	2	Production and purification of hydrogen from coke oven gas and other gases	Hydrogenation using coke oven gases.
	3	Dissolving coal and extracting constituents	Conversion of extraction constituents into high value materials, e.g. electrode cokes, carbon fibres.
	4	Hydrogenation and hydrocracking of coal, extraction products and tar aromatics	Production of hydrocarbons for chemical and plastics industries.
	5	Oxidation of coal, coal extracts and coal tar fractions	Production of chemicals containing carboxyl groups, etc.
	6	Absorption agents from coal	Activated carbons and coals for the purification of gaseous and liquid effluents.
	7	Electrode coke and reducing agents from coal	Special cokes with properties adapted to market requirements, e.g. for metallurgical and electrochemical industries
	8	Building and other materials from coal and shale	Cement from flotation waste, road-building materials, light-weight aggregates.
	9	Improved methods of combustion and utilisation of heat	New and improved combustion methods. Combined cycles for power generation. High pressure methods. Plasma physics.
	10	Microbiological treatment of coal	Production of protein.
	11	Technical and economical problems in environmental protection	New techniques and improvement of processes aimed at reducing noxious emissions.

APPENDIX D

MINES SAFETY AND HEALTH RECOMMENDATIONS
(13th Report, 1975) - EUROPEAN COMMISSION

A - Rescue

I - Consultation of foreign experts in the case of rescue operations connected with major mining accidents.

B - Fires and Underground Combustion

I - Equipment for shafts in connection with the prevention of fires.

II - Fighting of mine fires by sending down water.

III - Sealing off by stoppings of mine fires and underground combustion.

IV - Re-opening of sealed off fire areas.

V - Construction of advanced fire stoppings in plaster.

VI - Use of foamed urethane.

VII - Plaster stoppings.

VIII - Fire-resistant fluids.

IX - Tests and criteria for flammability for textile covered conveyor belts used in coal mines.

C - Electrification

I - Elimination of oil from underground electrical equipment.

II - Shotfiring leads.

III - Protection of underground distribution networks against the danger of causing electric shocks.

IV - Protection of underground electrical networks against fire and firedamp explosion risks.

V - Use of explosion-proof electrical apparatus for nominal voltage above 1 100 volts.

VI - Cables supplying mobile machines and their electrical protection.

D - Winding Ropes and Shaft Guides

I - Electro-magnetic examination of winding ropes.

II - Use of accelerometers to test winding installations.

E - Ventilation and Mine Gas

I - Practical conclusions on the application of the theory of stabilisation of ventilation.

II - Conditions under which exemption might be granted to raise maximum permitted CH_4 limits.

F - Mechanisation

I - Locomotive equipment.

II - Neutralisation of diesel engine exhaust fumes.

G - Health in Coal Mines and Medical Problems

I - Means of suppressing dust concentrations in underground workings.

II - Galleries and shafts.

III - Organisation of special services responsible for the inspection of dust conditions in underground workings.

IV - Fixing of climatic limits.

V - Medical examinations.

VI - Design and use of coal-getting and heading machines in relation to the reduction of airborne dust.

H - Psychological and Sociological Factors Affecting Safety

I - Psychological and sociological factors affecting safety.

II - Possible influence of payment at piece rates on safety in coal mines.

I - Flammable Dust

I - Procedure for binding dust by means of hygroscopic salts, another effective technique for neutralizing flammable dust.

II - Water barriers for containing dust explosions underground.

III - Triggered barriers.

APPENDIX E

ORGANISATIONS WITH MINING-ORIENTATED ACTIVITIES

Organisation	Field of Research	Mining Potential
Rutherford High Energy Laboratory	Facility for undertaking external research	General
Transport and Road Research Laboratory	Mechanics of cutting - use of discs	Cutting
	Pavement design - sandwich materials and epoxies	Floor quality
	Materials for roads and binding	Roadway stability
	Advanced steering and information display systems	Automatic/more reliable transport
Atomic Weapons Research Establishment	Peaceful uses of nuclear explosions	In-situ energy extraction
National Engineering Laboratory	Efficiency of pumps and turbines	Ventilation and hydraulics
	Materials and structure testing	Supports strength and fatigue
	Hydraulic impactors and drills	Rock breakage and tunnelling
	Heat engines and steam pumps	Pumping
Royal Armament Research and Development Establishment	Slurry explosives, shaped charges, linear cutting cord and sheet charges Plasma cutting	Rock breakage and cutting

Organisation	Field of Research	Mining Potential
Royal Signals and Radar Establishment (Malvern)	Infra-red imaging	Detection of heatings, over-heating machines, rescue work
	Laser techniques in spectroscopy and scattering	Gas analysis, air flow measurement
	Microprocessors	Distributed computational facilities
	Surface Acoustic Wave devices (SAW) and Charge Coupled Devices (CCD)	Signal processing and imaging
	Electro-luminescent displays	Large area, low power displays
Royal Signals and Radar Establishment (Baldock)	High power gas lasers	Cutting
	Compact medium power lasers	Survey and automatic alignment, gas analysis
	Infra-red imagers	Malvern
	Semi-conductor devices for efficient production of microwave power	Application of compact radar systems
Microbiological Research Establishment	Use of micro-organisms on coal	In-situ energy extraction
CEGB - Central Electricity Research Laboratories	Fibre optic data transmission and appropriate transducers	Sensing and transmission of data
	Robotics in fault diagnosis and repair	Remote operation
	Power station cooling, atmospheric pollution - relevance of fluidised bed combustion	Application of mining product. Importance of sulphur content
	Novel means of power generation	Affect on demand
	High power lasers	Cutting
	Energy transmission and storage	

Organisation	Field of Research	Mining Potential
Electricity Council	Cable jointing and finding	Distribution
	Plasma jet cutting	
	Heat pumps	Efficient drying of fine coal
	Compact batteries	Portable power
	Electrically powered vehicles	Transport
	Oxidation of atmospheric pollutant by ultra-violet light	Environment
National Physical Laboratory	Thermodynamic data bank	In-situ extraction process information
	Long term future of computer application	Data processing and control
	New materials - load bearing properties, strength, etc	Improved materials
	Corrosion	Failures of underground equipment
Culham Laboratory of Ukaea	Use of high power lasers in cutting, surface treatments and welding	Cutting and the production of good tool surfaces, bonds etc.
	Electrostatic ignitions in tankers	
Warren Spring Laboratory	Analysis of material handling hydraulic and pneumatic transport	Logistics and transport
	Microprocessors, colour displays	Control systems
	X-ray fluorescence spectrometers	Coal preparation
	Facility for rapidly undertaking external research problems	General
GEC - Hirst Research Centre	Microwave devices	Compact radar systems
	CCD and SAW devices	Compact low light, low power TV cameras, signal processing

Organisation	Field of Research	Mining Potential
	Fibre optics	Data transmission
	DC electro-luminescent displays	Large area, low power displays
	Teletext systems	Information supply
	Microprocessors	Data processing and control
BP- Sunbury Research Centre	Catalytic removal of pollutants	Environment
	Combustion of 'lean' gases and secondary recovery	In-situ extraction
	Computer control and information systems	Automation and control
Fulmer Research Institute	Hard coatings for refractory metals. High damping alloys and cutting tools. Surface treatment by high power lasers. Materials optimisation	General materials improvements
Wimpey Laboratories	Rock stabilisation	Strata control
	Prospecting	Exploration
	Drilling techniques	In-situ extraction
EMI - Central Research Laboratories	X-ray scanner and its fast data processing	Fast data processing
	Ultrasonic scanner and ultrasonic image converter	Position measurements in opaque atmospheres
	Speech recognition for computer input	"Hands off" data input
Plessey - Allen Clark Research Centre	Infra-red detectors	Thermal imaging
	Microwave equipment	Compact radars
	High density memories (magnetic bubbles and holograms)	Data processing
	Piezo-electric PVC	Vibration sensing
	Electrophoretic Displays	Large area, low power ambient light displays

Organisation	Field of Research	Mining Potential
	Fibre optics	Data transmission
Cementation - Rickmansworth Laboratories	Pump packing Solids pumping Drilling	Strata control mineral transport
	Hydrofracture	In-situ extraction
	Acoustic/vibratory techniques for fissure detection	In-seam seismology
Hawker Siddeley Dynamics Engineering Ltd.	Turbine control, automatic gearboxes for heavy haulage	Engine control
	Axle box overheat detector Present work with MRDE including underground monitoring and control of a variety of devices	Current work in monitoring and control
Dosco Overseas Engineering Ltd.	Mining machinery	
	Consideration of departures from conventional machines	(details not divulged at present)
Hawker Siddeley Dynamics Ltd.	Logistics	Colliery organisation
	Satellite and subseam equipment	Equipment for harsh environments
	Mass flow measurement systems	Mineral transport
	Closed loop servo systems	Control
Rio Tinto-Zinc Headquarters	Ore mining and surface coal gasification	General and in-situ extraction
Shell International, Shell Centre	Lagging techniques and drilling	Exploration
	Hydraulic and 'cavity' mining. In-situ extraction techniques	Non-conventional extraction techniques
	Surface processing	Form of product

Organisation	Field of Research	Mining Potential
Philips - Mullard Research Laboratories Redhill	Semiconductor devices, integrated circuit developments	More compact data processing
	Microwave systems, compact radar systems	Compact vehicle radar for guidance in murky atmospheres or remote control
	Telephone techniques and teletext services	Information techniques
	Computer character recognition and visual scene analysis	Computer controlled operations (robotic)
	Compact medium power lasers and infra-red techniques	Automatic alignment surveying gas analysis and thermal imaging
Davy International Group	Submarine dredging	Sea floor opencasting
	Instrumentation and hydraulics	General
ITT - Standard Telecommunications Laboratories (STL)	Optical fibre technology cables and repeaters	Data transmission
	Silicon technology	Electronics, strain gauges
	Ultrasonic techniques	Velocity measurements, communications
	Microprocessed sensors	More reliable transducers
	Non electrical switching	Intrinsic safety
	Liquid crystals, electroluminescence, LED, Beta lights	Displays
	Causes of machine breakdown	Machine reliability
Esso Research Laboratories, Abingdon	Bitumen and asphalt uses	Floor and strata stablisation
	Suppression of hazards from pollution and explosions	Environment
	Slurry transportation	Mineral transport
	Drilling techniques	

Organisation	Field of Research	Mining Potential
	Ash and sulphur removal	Coal preparation
	Application of new techniques	General
Vickers Ltd. & International Research and Development Ltd.	Motors and generators	Transport
	Microprocessors	Control
	Pump bearings	
	Lasers	Alignment
	Noise reduction	
ICI Corporate Research Laboratories	Computer control, information presentation. Choice of central or distributed computation	General computer/automation philosophy
	Robotics	Remote operation
	Pneumatic and hydraulic conveying	Mineral transport
	Material stabilisation	Strata control
	Surface phenomena on membranes	Gas detection
	Colloidal science	Slurry explosives and coal preparation
	Systems engineering	Integration of operations
	Process technology	
U.S.A.		
ERDA, Washington	Directional drilling of holes, rubblising of seams	Underground gasification
USBM, Washington	Advanced mining systems and a variety of in-situ methods including hydraulic mining and methane recovery	Non conventional mining methods
Universities of West Virginia and California	A wide range of techniques for in-situ extraction	Non conventional mining methods In-situ extraction

Organisation	Field of Research	Mining Potential
USBM, Twin Cities	Borehole hydraulic mining	In-situ techniques
GERMANY		
Technical High School Aachen	Cutting and drilling, ploughs, shearers, steering etc, ie work on face machinery in general	Face machinery
StBV, Essen Kray	Noise reduction	Noise
	"Coal Mine of the Future"	General
Rösler K.G. General Blumenthal & Prosper IV Mines	Use of "Drafonet" plastic covered wire netting for roof control	Roof control

APPENDIX F

PETROLEUM FEEDSTOCKS & SOME OF THEIR DERIVATIVES

Primary & Secondary Feedstocks	Main Product	End Product
methane:		
hydrogen	chemical intermediates	
	ammonia	fertilizers
hydrogen sulphide	sulphur	sulphuric acid
hydrogen cyanide	acrylonitrile	clear plastics
	acetone	chemical intermediate, solvent
chloromethane	carbon tetrachloride	degreaser
acetylene	chloroprene	polychloroprene for neoprene rubbers
	vinyl chloride	polyvinyl chloride (PVC) for plastics, adhesives, emulusion paint
	methyl methacrylate	clear plastics
	acrylic acid, acrylonitrile, chloroethane	
fluorethane	polytetraflurethane	safety glass, non-stick coatings
methanol	chemical intermediates	
	formaldehyde	pharmaceuticals, solvents
	methylamines	special solvents

Primary & Secondary Feedstocks	Main Product	End Product
syngas	synthetic ammonia methanol (see above)	
ethane:		
acetylene	see methane	
chloroethane	ethylbenzene see benzene	
nitroethane	chemical intermediates	
ethanol	ethylamines	solvents, methylated spirit
	chlorobenzene	pharmaceuticals, surface acting agents
ethylene	acetaldehyde	acetic acid, acrylic acid, and formic acid for cellulose plastics
	acetylene see methane	
	polyethylene	plastics
	chloroethylene (ethylene dichloride)	vinyl chloride for plastics
		fuel additive (TEL) (dichloroethane)
		adhesives, degreaser (trichloroethane)
	ethyl dibromide	fuel additive (TEL)
	ethylene oxide	detergents, explosives, solvents, epoxy resins
		anti-freeze (ethylene glycol)
		solvent (diethylene glycol)
		solvent (triethyleneglycol)
	ethylbenzene see benzene	
	ethanol (see above)	
propane:	nitric acid	chemical intermediate

Primary & Secondary Feedstocks	Main Product	End Product
	nitropropane	chemical intermediate
	nitromethane	chemical intermediate
propylene	cumene	
	cumene hydroperoxide	acetone for paints, brake fluid polymethyl methacrylate for clear plastics
		phenol for pharmaceuticals plastics, explosives
	acrylonitrile see methane	
	propylene oxide	detergents, paints polyurethane foam
	isopropanol	acetone for cosmetics, paints
	acrylic acide	acrylates for textile finishes, adhesives
	polypropylene butyraldehydes	film, plastics
propane and butane:		
ethylbenzene	see benzene	
formaldehyde	see benzene	plastics, dyes, disinfectants
acetaldehyde	see ethylene	
methanol	see methane	
acetic acid	cellulose	plastics
butane:		
butadiene	acrylonitrile see methane	nitrile rubber
	polychloroprene	neoprene rubber
	with styrene	nylon, synthetic rubber
butylene	isobutylene	butyl rubber, sealing compounds, oil additive
butan-2-ol	butanone	solvent

APPENDIX G

SOURCES OF INFORMATION - COAL PRODUCING COUNTRIES

AUSTRALIA:	Queensland Coal Board, 169 Mary Street, Brisbane.
	Department of Mines, New South Wales.
	The CAGA Centre, 8 Bent Street, Sydney 2000.
	Ministry of Fuel & Power-Victoria, 151 Flinders Street, Melbourne 3000.
BELGIUM:	Iniex, rue du Chera 200, Liege, Belgium 4000.
CANADA:	Department of Energy, Mines and Resources, 555 Booth Street, Ottawa.
CHINA:	Coal Mining Ministry, Peking.
FRANCE:	Charbonnages de France, 9 Av. Percier, 75008 Paris.
GERMANY:	Bundesministerium für Wirtschaft, 5300 Bonn-Duisdorf, Villemombler Str. 76. (Unterabteilung III is concerned with mining)
	Steinkohlenbergbauverein, 4300 Essen-Kray, Frillendorfer Str. 351.
	Ruhrkohle AG, 4300 Essen 1, Postfach 5.
INDIA:	India Coal Owners Association, U19 Green Park Extension, New Delhi 16.
	Coal India, 10 North Sabhas Road, Calcutta 700001.

POLAND:	Ministry of Mining, Krucza 36, 062921 Warsaw.
SOUTH AFRICA:	Chamber of Mines, P.O. Box 809, Johannesburg 2000.
UNITED STATES:	US Bureau of Mines, 2401 E Street NW, Washington DC 20241.
U.S.S.R.	Ministry of the Coal Industry of the U.S.S.R., 23 Kalinin Avenue, Moscow.
U.K.	National Coal Board, London S.W.1.

APPENDIX H

GERMAN INDUSTRIAL INVOLVEMENT IN COAL RESEARCH AND DEVELOPMENT

Company	Project
Arbeitsgemeinschaft Formkoks	Production of formed coke (demonstration plant Prosper)
Arbeitsgemeinschaft Ruhrgas AG, Ruhrkohle AG, STEAG AG, Essen	Production of synthesis gas/SNG/town gas by the LURGI pressure gasification
Arbeitsgemeinschaft Saarbergwerke AG, Saarbrücken	Technology centre for pressure gasification of pulverised coal: Development of processes for the gasification of coal under pressure
Arbeitsgemeinschaft Wirbelschicht-feuerung	Preparation of a data base for a 100 MW coal-fired power station ready for construction using fluidised bed combustion technology (main phase)
Atlas-Copco	Improved performance from use of large caliber safety charges and flameproof detonator cord in gates and gate-end roads
BAG	Developing ways of using structural data from tectonics in coalmining
	Development of reflection seismic process for revealing carboniferous strata, to remain effective even with the prevalence of permian formations, in particular rock salt
	Steering and partial automation of winning installation
BASF AG, Ludwigshafen	Enlargement of the raw material basis for refineries by inclusion of hard coal (laboratory phase)
Battelle-Institut e.V. Frankfurt/Main	Coal mining of the future - extension of mining techniques sub-project 7. Conveying techniques

Company	Project
	Coal mining of the future - extension of mining techniques sub-project 10. Operations management and engineering
Becorit/Gullik	Testing different systems of shield-type supports
	Shield support for thick seams with extremely poor roof conditions
	Caving screen for use with powered support systems
Bergbau-Forschung GmbH, Essen	Preliminary degasification using the LR process
	Technology centre for pressure gasification of pulverized coal: Development of processes for the gasification of coal under pressure
	Production of synthesis gas/SNG/town gas by the LURGI pressure gasification process (Dorsten project)
	Development of processes for coal gasification with nuclear process heat (immersion heater process)
	Working material for the gasification of coal according to the immersion heater process (development)
	Preparation of a data base for a 100 MW coal-fired power station ready for construction using fluidized bed combustion technology (main phase)
	Thermal pre-treatment of baking pit coal
	Production of formed coke (demonstration plant Prosper)
	Development and testing of flameproof television equipment for photographing seams and adjoining rock and tectonic surfaces in underground boreholes
	Development of improved methods of using geophysical borehole surveys for locating carboniferous strata
	Reconnaissance of area immediately in front of face through long-distance horizontal drillings

Company	Project
Bergbau - Forschung GmbH, Essen	Verifying possibility of using counter-flush process for underground reconnaissance of the area immediately in front of the face
	Developing ways of using structural data from tectonics in coalmining
	Prospecting for new deposits by means of exploratory borings from the surface
	Development of reflection seismic process for revealing carboniferous strata, to remain effective even with the prevalence of permian formations, in particular rock salt
	Study of characterisation and quantitative logging of feasibility of coal
	Further development of automatic measuring equipment for X-ray mineral analysis by computer
	Improved performance from use of large calibre safety charges and flameproof detonator cord in gates and gate-end roads
	Improvements of designs and techniques of winning with ploughs
	New winning system through simultaneous cutting and crushing
	Development of a drive unit for chain scraper face-conveyors with defined chain pre-tension
	Research on basic principles applying to process control of coal preparation
	Building and integration of operational planning models for the coalmining industry in the Federal Republic of Germany (short title: Mining planning models) Sub-Project 1
	Building and integration of operational planning models for the coalming industry of the Federal Republic of Germany (short title: Mining Planning Models) Sub--Project 2

Company	Project
Bergbau - Forschung GmbH, Essen	Building and integration of operational planning models for the coal mining industry of the Federal Republic of Germany (short title: Mining Planning models) Sub-Project 3
	Early diagnosis and prevention of mining damage
	Coal mining of the future - extension of mining techniques sub-project 1. Basic principles and co-ordination
	Coal mining of the future - extension of mining techniques sub-project 2. Layout
	Coal mining of the future - extension of mining techniques sub-project 3. Deposits
	Coal mining of the future - extension of mining techniques. Sub-project 4. Rock mechanics
	Coal mining of the future - extension of mining techniques sub-project 5. Drifting and expansion of underground workings
	Coal mining of the future - extension of mining techniques sub-project 6. Methods of coal mining.
	Coal mining of the future - extension of mining techniques sub-project 7. Conveying techniques.
	Coal mining of the future - extension of mining techniques sub-project 8. Mine climate
	Coal mining of the future - extension of mining techniques sub-project 9. Surface installations
	Coal mining of the future - extension of mining techniques sub-project 10. Operations management and engineering
	Study regarding technologies for rational use of energy
C.F. Braun & Co., USA	Manufacture of petrochemical raw materials from synthesis gas (study)
Braunschweigische Kohlen Bergwerke	Economical production and utilisation of Posidonian shale from Schandelah

Company	Project
Brieden & Co.	Rationalisation of support process through improved backfilling techniques
	Construction of gateside packs to increase creeping strength in gate-roads
	Boreshaft with hydraulic haulage system
Burderus'sche Eisenwerke, Dietzhölztal	The increase of the utilisation of the primary energy and reduction of the emission with fully automatic, grate-fired, solid fuel heat generators
Bundesanstalt für Geowissenschaften u. Rohstoffe Hannover	Development of organic-geochemical and isotope geophysical methods for use in the exploration for hydrocarbons
BWD, Rheinland	Development and testing of flameproof television equipment for photographing seams and adjoining rock and tectonic surfaces in underground boreholes
Cherchar Industries	Steering and partial automation of wining installation
Deilmann-Haniel	Improved performance from use of large caliber safety charges and flameproof detonator cord in gates and gate-end roads
	Improved performance, easier working and increased safety of support operations in drifting
Demag	Re-development of road heading machines
	Fully mechanised driving of gate-end roads in the rock conditions peculiar to the Aachen anthracite field
Deutscher Braunkohlen Industrieverein, Köln	Exploration of brown coal deposits by use of secondary indicators recognised by remote sensing method
Didier GmbH, Essen	Methanisation of coal gasification gases in fluidized bed
Gebr. Eickhoff GmbH	Eickhoff EVA 160 heading machine with in-built dust suppression equipment and support erection device
	Trial of chainless haulage systems for shearer loaders
	Cutting a wide section of a gateroad kept on the face line by an additional shearer loader

Company	Project
Eisenhütte Westfalia	Development of a drive unit for chain scraper face-conveyors with defined chain pre-tension
Eschweiler Bergwerksverein	Fully mechanised driving of gate-end roads in the rock conditions peculiar to the Aachen anthracite field
	Integrated coal face support equipment for very thick seams
	Re-development on high performance screening machines for pre-screening
	Development of clarifying facilities and processes
Herzogenrath-Kohlscheid	Building and integration of operational planning models for the coalmining industry in the Federal Republic of Germany (short title: Mining planning models) sub-project 1
Geologisches Landesamt NRW	Coal mining of the future - extension of mining techniques sub-project 3. Deposits
Gesellschaft für Vergasung und Verflüssigung, Essen	Demonstration plant for the production of coal oil in the USA
GHH Sterkrade AG, Oberhausen	Preliminary studies: Extra-high pressure water jet coal-winning machine
Gluckauf Bau GmbH	Bowl centrifuge for dewatering froth tailings
Halliburton	Horizontal and vertical hydraulic conveyance of abrasive solids from great depths
Hausherr	Shield-type supports and impact ripper machines in the transition from coal face to roadway
Hemscheidt, Wuppertal-E	Improved drift advance with road heading machines by means of support-erecting aids
	Testing different systems of shield-type supports
Ibak, Kiel	Development and testing of flameproof television equipment for photographing seams and adjoining rock and tectonic surfaces in underground boreholes

Company	Project
IIT Research Institute Chicago, USA	Preliminary studies: Extra-high pressure water jet coal-winning machine
Ingenieurbüro Meissner Ebert & Partner, Nürnberg	Replacement of light fuel oil in the range of central heating by pulverized coal
Ingersoll	Improvement of drilling and blasting techniques in gate-end roads
Itak, Celle	Prospecting for new deposits by means of exploratory borings from the surface
Kammerich/Reisholz GmbH, Düsseldorf	Prototype plant nuclear process heat
Kernforschungsanlage Jülich GmbH	Scientific investigations into the utilization of coal
	Fluidized combustion under pressure
	Data bank of world coal resources and reserves
	Technical information service
	Mining technology clearing house
KHD Humboldt Wedag	Steam cap filter for dewatering froth fines
	Bowl centrifuge for dewatering of froth tailings
	Process for the treatment of coals preliminarily crushed and difficult to prepare and for a reduction of the water and ash contents of coal
KDH Industrieanlagen AG, Köln	Particle analyser for on-line-monitoring in coal preparation
Klöckner-Ferromatik	Development of an additional device for securing the coalface in conjunction with shield supports with thick seam working
	Transfer system between the coal face to roadway conveyor and the high capacity crusher
Koppers, Krupp, Essen	Coal gasification in the tube-furnace gasifier (pilot plant)
Krauss-Maffei-Imperial München	Pusher centrifuge for reprocessing of filter cakes

Company	Project
Küppers & Röllecke	Process for the treatment of coals preliminarily crushed and difficult to prepare and for a reduction of the water and ash contents of coal
Lurgi, Mineralöltechnik	Preliminary degasification using the LR process (study)
	Production of synthesis gas/SNG/town gas by the LURGI pressure gasification process (Dorsten project)
	Prototype-plant for nuclear process heat conception-phase of the project
	Manufacture of petrochemical raw materials from synthesis gas (study)
	Synthesis of raw materials for the chemical industry by means of the Fischer-Tropsch process (study, simulation programme)
	Development of technical processes for the treatment of tar from coal conversion
	Further development of a 170 MW prototype plant at the Lünen power station operating with coal pressure gasification system and combined gas/steam turbine cycle process
Mannesmann-Forschungsinstitut, Duisburg	Working material for the gasification of coal according to the immersion heater process (development)
Messerschmidt-Bölkow-Blohm, München	Improvement of drilling and blasting techniques in gate-end roads
	Coal mining of the future - extension of mining techniques sub-project 10. Operations management and engineering
	Technological research on the development of an extra-high pressure water jet pulsator gun
Mitgliedsges. des Stbv	Coal mining of the future - extension of mining techniques sub-project 3. Deposits
Neuhaus	Improved performance, easier working and increased safety of support operations in drifting

Company	Project
Paurat	Re-development of road heading machines
Prakla-Seismos, GmbH, Hannover	Prospecting for new deposits by means of exploratory borings from the surface
	Development of reflection seismic process for revealing carboniferous strata, to remain effective even with the prevail- ence of permian formations, in particu- lar rock salt
Preussag AG, Ibbenbüren	The increase of the utilization of the primary energy and reduction of the emission with fully automatic, grate- fired, solid fuel heat generators
	Process for the treatment of coals pre- liminary crushed and difficult to prepare and for a reduction of the water and ash contents of coal
Rheinische Braunkohlenwerke AG, Köln	Coal gasification in the high-temperature Winkler gasifier (pilot plant)
	Coal gasification in the tube-furnace gasifier (pilot plant)
	Development of processes for the con- version of coal by using heat from gas cooled high-temperature nuclear reactors (hydrogasification)
	Prototype-plant for nuclear process heat conception phase of the project
Rheinstahl AG	Combustion of tailings with fluidized- bed firing
Ruhrchemie AG	Methanization of coal gasification gases in fluidized bed
	Development of catalysts for the Fischer- Tropsch-Process
Ruhrkohle AG, Essen	Production of a synthesis gas by pres- sure gasification of coal dust with oxygen and water according to the Texaco Gasification Process
	Demonstration plant for the production of coal oil in the USA
	Production of formed coke (demonstration plant Prosper)
	Prospecting for new deposits by means of exploratory borings from the surface

Company	Project
Ruhrkohle AG, Essen	Improvement of drilling and blasting techniques in gate-end roads
	Improved performance from use of large caliber safety charges and flameproof detonator cord in gates and gate-end roads
	Further development of rotary dry drilling for soft rock conditions
	Development and testing of a non compressed air high-performance drivage system with drilling and blasting
	High-performance drivage by shot firing
	Re-development of road heading machines
	Improved drift advance with road heading machines by means of support-erecting aids
	Further development work on rock tunnelling on the dip with the Full Face Tunnelling Machine system with hydraulic debris disposal
	Support with precast concrete elements (panel-type support)
	Technological and conveying methods for the use of water-setting cementing materials
	Rationalisation of support process through improved backfilling techniques
	Construction of gateside packs to increase creeping strength in gate-roads
	Tests with new types of lagging
	Automated shaft sinking with the use of a staple-shaft drilling machine, developing a technique for diameters drilled in excess of 5 m
	Boreshaft with hydraulic haulage system
	Improvements of designs and techniques of winning with ploughs
	Steering and partial automation of winning installation

Company	Project
Ruhrkohle AG, Essen	New winning system through simultaneous cutting and crushing
	Development of a drive unit for chain scraper face-conveyors with defined chain pre-tension
	Development of a crusher installed in the face conveyor system
	Trial of chainless haulage systems for shearer loaders
	Testing different systems of shield-type supports
	Development of an additional device for securing the coalface in conjunction with shield supports with thick seam working
	Caving screen for use with powered support systems
	Integrated coal face support equipment for very thick seams
	Supports for the transition between coal face and roadway
	Shield-type supports and impact ripper machines in the transition from coal face to roadway
	Transfer system between the coal face to roadway conveyor and the high-capacity crusher
	Cutting a wide section of a gateroad kept on the face line by an additional shearer loader
	Optimization of the sequence of operations at the junction of gateroad with a coal face worked by coal ploughs
	Mining in inclined deposits
	Mineworking with detonator cords
	Hydropit in the Ruhr deposits
	Battery driven suspension monorail locomotive
	Trackless, diesel-operated self-dumping transport wagon (shuttle car)

Company	Project
Ruhrkohle AG, Essen	Transport facilities in roadways by means of ski-lift type installations
	Diesel engines for operation in high CH_4 content
	Employment of batteries without plate protection
	Development of geared drive systems for ground mounted rail conveyor and suspension monorail conveyor
	Computer controlled optimization of train movements
	10 kV supply for medium voltage network
	Selective earth-fault monitoring in medium voltage network
	Power factor improvement in underground power supply network
	Transmission of starting characteristics
	Employment of 5 kV motors and fast connections for roadway conveyor system drives
	Development of power lines of greater flexibility for cutting machinery
	Mobile, flameproof power leads
	Development of on-load double-throw disconnecting switches and use of vacuum contactors
	Intercommunications system for underground working
	Monitoring, evaluation and storage of measured values and other data on underground mineworking with the use of a process computer
	Steam-cap filter for dewatering froth fines
	Bowl centrifuge for dewatering of froth tailings
	Process for the treatment of coals preliminarily crushed and difficult to prepare and for a reduction of the water and ash contents of coal

Company	Project
Ruhrkohle AG, Essen	Practical testing of the selective agglomeration of coal
	Combustion of tailings with fluidized bed firing
	Research on basic principles applying to process control of coal preparation
	Project study on process engineering tie-up between homogenisation of raw coal and end products
	Computer-controlled coal preparation
	Monitoring and control of conveyance flow in underground working
	Homogenisation of feed to the washery
	Building and integration of operational planning models for the coal mining industry of the Federal Republic of Germany (short title: Mining planning models) sub-project 3
	Coal mining of the future - extension of mining techniques. Sub-project 4 Rock mechanics
	Coal mining of the future - extension of mining techniques. Sub-project 9. Surface installations
RWTH, Aachen	Coal gasification in the high-temperature Winkler gasifier (pilot plant)
	Coal gasification in the tube-furnace gasifier (pilot plant)
	Pressure change-underground gasification
	Development of homogeneous catalysts for coal hygrogenation
	Coal mining of the future - extension of mining techniques sub-project 6. Methods of coal mining
	Coal mining of the future - extension of mining techniques sub-project 7. Conveying techniques
	Coal mining of the future - extension of mining techniques sub-project 8. Mine climate

Company	Project
Saarberg Interplan	Preparation and conception of a field test for underground gasification of bituminous coal
	Production of methane from coal bearing formations by drilling from surface
	Prospecting for new deposits by means of exploratory borings from the surface
Saarbergwerke AG, Saarbrücken	Preparation and conception of a field test for underground gasification.
	Enlargement of the raw material basis for refineries by inclusion of hard coal (laboratory phase)
	Enlargement of the raw material basis for refineries by inclusion of hard coal (accompanying study to the laboratory phase)
	Preparation of a project study for the erection of a demonstration plant for desulphurisation of stack gas using the Saarberg-Hölter process
	Prospecting for new deposits by means of exploratory borings from the surface
	Production of methane from coal bearing formations by drilling from surface
	Eickhoff EVA 160 heading machine with in-built dust-suppresion equipment and support-erection device
	Development of shearer loader for coal extraction from a seam 1 m thick
	Improvements to conventional powered supports in order to increase their operating width
	Supporting methods in coal faces, worked by pneumatic stowing, in thick seams with difficult roof conditions
	Improvement of techniques for rock consolidation and for securing the roof
	Improvements to control techniques and automation with belt conveyors
	Signal recording in personnel-conveying
	Pusher centrifuge for reprocessing of filter cakes

Company	Project
Saarbergwerke AG, Saarbrücken	Jig for simultaneous cleaning of two different types of coal needing to be kept apart
	Research on basic principles applying to process control of coal preparation
	Building and integration of operational planning models for the coalmining industry of the Federal Republic of Germany (short title: Mining Planning Models) sub-project 2
	Coal mining of the future - extension of mining techniques sub-project 7. Conveying techniques
Salzgitter AG, Salzgitter	Further development of rotary dry drilling for soft rock conditions
	Development and testing of a non-compressed air high performance drivage system with drilling and blasting
SASOL, Südafrika	Manufacture of petrochemical raw materials from synthesis gas (study)
Dr. Schäfer, Essen	Monitoring and control of conveyance flow in underground working
Schering AG, Bergkamen	Synthesis of raw materials for the chemical industry by means of the Fischer-Tropsch process (study, simulation programme)
	Erection and operation of a Fischer-Tropsch pilot plant with liquid phase (slurry) reactor
Schlumberger	Development of improved methods of using geophysical borehole surveys for locating carboniferous strata
	Prospecting for new deposits by means of exploratory borings from the surface
Schmidt, Kranz	Mineworking with detonator cords
Siebtechnik GmbH Mülheim/Ruhr	Big centrifuge plant for smalls under 6 mm with a high content of fines
	Pusher centrifuge for reprocessing of filter cakes
Siemag GmbH	Further development work on rock tunelling on the dip with the Full Face Tunnelling Machine system with hydraulic debris disposal

Company	Project
Siemag GmbH	Boreshaft with hydraulic haulage system
	Hydropit in the Ruhr deposits
Siemens AG, Erlangen	Steering and partial automation of winning installation
	Employment of 5 kV motors and fast connections for roadway conveyor system drives
STEAG, Essen	Preliminary degasification using the LR process (study)
	Production of synthesis gas/SNG/town gas by the LURGI pressure gasification process (Dorsten project)
	Prototype plant nuclear process heat
	Demonstration plant for the production of coal oil in the USA
	Further development of a 170 MW prototype plant at the Lünen power station operating with coal pressure gasification system combined gas/steam turbine cycle process
L+C Steinmüller Gummersbach	VEW coal conversion process (pilot plant 1 t/h)
	VEW coal conversion process (study for 15 t/h plant)
Still, Recklinghausen/Essen	VEW coal conversion process (study for 15 t/h plant)
	Thermal pre-treatment of baking pit coal
	Production of formed coke (demonstration plant Prosper)
Thyssengas GmbH, Duisburg	Methanization of coal gasification gases in fluidized bed
Thyssen-Schachtbau Mülheim/Ruhr	Verifying possibility of using counter flush process for underground reconnaissance of the area immediately in front of the face
	Prospecting for new deposits by means of exploratory borings from the surface
TU Berlin	Basic research for the selectibe guidance of the Fischer-Tropsch synthesis

Company	Project
TU Berlin	Physical-chemical investigations of bituminous coal extracts and their fractionation
	Coal mining of the future - extension of mining techniques sub-project 2. Layout
	Coal mining of the future - extension of mining techniques sub-project 3. Deposits
	Coal mining of the future - extension of mining techniques sub-project 5. Drifting and expansion of underground workings
TU Braunschweig	Calculation of the process dynamics of the start-up for a coal-fired power plant using fluidized combustion and combined cycles
TU Clausthal	Development of improved methods of using geophysical borehole surveys for locating carboniferous strata
	Developing ways of using structural data from tectonics in coalmining
	Particle-analyser for on-line-monitoring in coal preparation
	Coal mining of the future - extension of mining techniques sub-project 1. Basic principles and co-ordination
	Coal mining of the future - extension of mining techniques sub-project 2. Layout
	Coal mining of the future - extension of mining techniques sub-project 6. Methods of coal mining
	Coal mining of the future - extension of mining techniques sub-project 7. Conveying techniques
	Coal mining of the future - extension of mining techniques sub-project 10. Operations management and engineering.
TU Darmstadt	Synthesis of olefins in the range C_3 - C_{20} by Fischer-Tropsch liquid phase
Friedrich Uhde GmbH Dortmund	Coal gasification in the hightemperature Winkler gasifier (pilot plant)
	Manufacture of petrochemical raw materials from synthesis gas (study)

Company	Project
Friedrich Uhde GmbH Dortmund	VEW coal conversion process (study for 15 t/h plant)
Union Rheinische Braunkohlen Kraftstoff, Wesseling	Manufacture of petrochemical raw materials from synthesis gas (study)
Uni Bochum	Coal mining of the future - extension of mining techniques sub-project 3. Deposits
Uni Bonn	Basic research on catalysts for carbon monoxide hydrogenation
	Coal mining of the future - extension of mining techniques sub-project 3. Deposits
Uni Dortmund	High temperature gas-solid suspension cooling in heat exchangers (accompanying study to the VEW process)
Uni Erlangen-Nürnberg	Methanisation of coal gasification gases in fluidized bed
Uni Frankfurt	Developing ways of using structural data from tectonics in coalmining
Uni Karlsruhe	Investigation of the reaction kinetics of carbonisation and partial gasification of preoxidized coal powders (accompanying investigations to the VEW process)
	Simultaneous conversion and methanisation of CO-rich gases
	Catalysts and control of selectivity in the Fischer-Tropsch-Synthesis system
	Coal mining of the future - extension of mining techniques sub-project 4. Rock mechanics
Uni Kiel	Developing ways of using structural data from tectonics in coalmining
Uni München	Newly styled and improved catalysts for coal hydrogenation
Verbundwerk Haus Aden	Computer-controlled coal preparation
Vereinigte Elektrizitätswerke Westfalen AG, Dortmund	VEW coal conversion process (pilot plant 1 t/h)
	VEW coal conversion process (study for 15 t/h plant)

Company	Project
VGB, Essen	Distribution of sulphur in intermediate and residual coke investigations connected with the VEW process)
Voest-Alpine	Re-development of road heading machines
Westfalie, Lünen	Re-development of road heading machines
	Testing different systems of shield-type supports
	Shield-type supports and impact ripper machines in the transition from coal face to roadway
Winkhaus & Patschul, Essen	Coal mining of the future - extension of mining techniques sub-project 9. Surface installations
Zentralstelle für Geo-Photogrammetrie der Aug. Geo. Uni München	Developing ways of using structural data from tectonics in coalmining

APPENDIX I

SUPPLIERS OF MINING EQUIPMENT

Key

1. Agitators, Conditioners and Mixers
2. Analyzers
3. Belts and Belting
4. Blasting Supplies
5. Blending Machines
6. Breakers
7. Buckets and Teeth
8. Cars, Mine
9. Centrifuges
10. Chain Hoists
11. Coal Cutters
12. Coal Preparation Plants
13. Communications
14. Compactors
15. Compressors, Air and Accessories
16. Concentrating Equipment
17. Consulting Engineers
18. Continuous Miners
19. Controls
20. Conveyors and Elevators

21. Crushers

22. Cyclones

23. Drills, Exploration

24. Drills, Mining

25. Drills, Rock Big Hole

26. Dryers and Kilns

27. Dust Collection Equipment

28. Dust Sampling Equipment

29. Dust Suppression Equipment

30. Environment Protection Equipment

31. Environmental Quality and Measuring Services

32. Excavators

33. Exploration Equipment

34. Exploration Services

35. Filter Media

36. Filters

37. Flotation Machines

38. Flotation Reagents

39. Grinding Equipment

40. Hammers, Air Powered, Rock Breaking and Demolition

41. Heading Machines

42. Hoisting Equipment (Winders)

43. Hydraulic Power Packs

44. Lighting

45. Loaders, Front End and Overhead, and Fork Lift

46. Long Wall Miners

47. Mine Design

48. Personnel Carriers

49. Precipitators, Electrostatic

50. Props

51. Pulverizers
52. Pumps
53. Recorders
54. Roof Support
55. Safety Equipment
56. Self-Loading Transport
57. Separators
58. Shaft Sinking
59. Short Wall Miners
60. Thickeners and Tanks
61. Undercutters
62. Ventilation Equipment and Blowers
63. Washers
64. Water and Waste Treatment

Australia

Australian Mineral Development Lab., Flemington St., Frewville, S.A. 5063	2, 17, 19, 31
Fox Manufacturing Co., Div. Placer Explorations Ltd., 106-128 Woodpark Rd., Box 34, Smithfield, N.S.W.	8, 12, 17, 18, 19, 20, 24, 28, 34, 42, 48, 53.
Golder Associates (Pty) Ltd., 466 Malvern Rd., Prahan (Melbourne), Vic. 3181.	17, 47
Mindrill Ltd., 36-40 Northern Rd., Heidelberg West, Vic. 3081.	23, 34
Mineral Deposits Ltd., 81 Ashmore Rd., Box 44, Southport, Qnsld. 4215.	16, 17, 31, 34, 57
Mitchell Cotts Project (Australia) Pty Ltd., 246 St. George's Terrace, Perth, W. Australia.	12, 16, 17
Moss (Pty) Ltd., George, Box 106, Mt. Hawthorn WA 6016.	8, 23, 24, 62
Parbury Henry & Co., Mining Div. 1 Lincoln St., Lane Cove West, Sydney, N.S.W.	47,
Production Equipment Ltd., Fairbank Rd., Clayton,Vic. 3168	12, 17, 20, 32, 42

Titan Mfg. Co., Box 292, Newcastle, N.S.W.	20, 48, 50, 54, 55
Vickers Australia Ltd., Vickers Ruwolt Div., 524 Victoria St., Richmond, Vic. 3121.	6, 7, 11, 12, 16, 17, 20, 21, 22, 26, 36, 39, 45, 46, 52
Warman Equipment (International) Ltd. 4 Marden St., Atarmon, Sydney, N.S.W. 2064.	1, 4, 12, 16, 17, 20, 22, 27, 36, 37, 52, 63, 64

Austria

Bohler Bros. & Co., Ltd., Niebelungengasse 8, A-1010 Vienna.	6, 24, 52
Rheax Chemie & Metall GmbH, Box 578, Wollzeile 12, A-1011 Vienna.	12, 16, 17, 22, 57, 60, 63, 64
Schaffler & Co., Sturzgasse 34, A-1150 Vienna.	4,
Vereinigte Osterreichische Alpine Montan, Eisen & Stahlwerke, Prinz-Eugen Str. 8-10, A-1040 Vienna.	12, 20, 29, 41, 42, 50, 52, 54.
Voest-Alpine, Postf. 2, A-4010 Linz/Donau.	12, 17, 18, 20, 21, 26, 27, 36, 47, 50, 52, 58, 60, 64.

Belgium

Birtley NV, SA, 32-36 Ave. de Tervueren, B-1040 Brussels.	9, 12, 16, 17, 21, 23, 26, 29, 36, 57
Clark International Marketing SA, Clark Bobcat Div., 43 Zevenbronnenstr., B-1512 Dworp.	32, 45
Colmant & Cuvelier SA, Blvd., Des Combattants, B-7500 Tournai.	3,
Joy International SA, 115 rue Defacoz, B-1050 Brussels	1, 6, 11, 15, 16, 18, 20, 21, 22, 24, 25, 27, 28, 29, 37, 39, 42, 45, 48, 49, 52, 56, 60, 62

Brazil

Gates Do Brasil SA, 602/634 Rua Cesario Alvim, CEP 03054 Sao Paulo, Brazil.	3, 20

Canada

AAF Ltd., 400 Stinson Blvd., Montreal, Que. H4N 261.	12, 22, 27, 30, 35, 36, 49, 57
Aerofalls Mills Ltd., 2640 S. Sheridan Way, Clarkson, Ont. L5J 2M8.	39,
Drill Systems Inc., 616 S.E. 58th Ave., Calgary, Alta. T2H OP8.	23, 24

Exploranium Corp. of Canada, Div. Geometric Services Ltd., 436 Limestone Cres., Downsview, Toronto, Ont. M3J 2S4.	33,
Gates Rubber of Canada Ltd., 50 Iroquois St., Brantford, Ont. N3T 5R6.	3,
Geonics Ltd., 51-2 Thorncliffe Park Dr., Toronto, Ont. M4H 1H2.	33,
Golder & Associates Ltd., H.Q., 3151 Wharton Way, Mississauga, Ont. L4X 2B6.	17, 47
Golder, Brawner & Associates Ltd., 224 W. 8th Ave., Vancouver, BC V5Y 1N5.	17,
Hepburn Ltd., John T., 914 du Pont St., Toronto, Ont. M6H 1Z2.	42, 55
Huntec ("70") Ltd., 25 Howden Rd., Scarborough, Ont. M1R 5A6.	33, 34
McPhar Geophysics Ltd. 55 Tempo Ave., Willowdale, Ont. M2H 2R9.	17, 33, 34, 53
Sandvik Canadian Ltd., Wear Parts Div., 6835 Century Ave., Mississauga, Ont. L5N 2L2.	7
Scintrex Ltd., 222 Snidercroft Rd., Concord, Ont. L4K 1B5.	2, 17, 33, 34
Shaw-Almex Ltd., Box 430,Parry Sound, Ont. P2A 2X4.	3
Simons-McBean Ltd., 7220 NE Fisher St., Ste. 440, Calgary, Alta T2H 2H8.	17,
Simpson Engineering Ltd., 5740 Yonge St., Toronto, Ont. M2M 3T4.	17,
Smit International Ltd., J.K., 81 Tycos Rd., Toronto, Ont. M6B 1W5.	23, 24
Teledyne Canada Mining Products, Box 130, Thornbury, Ont. NOH 2PO.	6, 24, 48
Wesdril Equipment, 2800 Viking Way, Richmond, BC V6V 1N4.	23, 53

Chile

ARMCO Chile S.A.1, Casilla 68-C, Concepcion.	39

Czechoslovakia

Pragoinvest, Ceskomoravska 23, 18056 Praha 9.	15, 21

Denmark

Niro Atomizer Ltd., 305 Gladsazevej, DK-2860 Soeborg.	26

England

Allens of Tipton Ltd., Box 4, Tipton West Midlands DY4 9EX.	8, 42, 58
Appleton & Howard Ltd., Bremar House, 27 Sale Pl., London W2 1PR.	52,
APV-Mitchell (Dryers), Ltd., Denton Holme Carlisle, Cumberland CA2 5DU.	1, 26
Aveling-Barford Ltd., Invicta Works, Grantham, Lincs.	21, 26, 45, 52, 63
Babcock-Moxey Ltd., Bristol Rd., Gloucester GL1 5RX.	5, 20, 32
Babcock & Wilcox Ltd., 165 Great Dover St., London SE1 4YB.	3, 5, 11, 12, 18, 19, 20, 53
Becorit (GB) Ltd., 2 Leslie Rd., Gregory Blvd., Nottingham.	48, 50, 55, 62
Beryllium Smelting Co. Ltd., 36138 Southampton St., London WC2E 7HJ.	39,
Boart (UK) Ltd., 14-18 High Holborn, London WC1V 6BX.	23, 24, 34
Boxmag-Rapid Ltd., Chester St., Aston, Birmingham B6 4AJ.	16, 20, 57
Bpb Industries Instruments Ltd., East Leake, Loughborough, Leics.	2, 19, 33, 34
British Jeffrey Diamond, Thornes Works Wakefield WF2 8PT.	6, 11, 21, 46, 51
British Ropeway Engineering Co. Ltd. Tubs Hill House, London Rd., Seven Oaks, Kent TN13 1DB.	20, 31
BTR Belting Ltd., Box 3, Centurion Way, Leyland, Lancs PR5 2RE.	3,
Cable Belt Ltd., 15 Victoria Ave., Camberley, Surrey GU15 3HS.	3, 17
Cambridge Instruments Ltd. Melbourn, Royston, Herts SG8 6EJ.	2,
Celtite Selfix Ltd., Rough Close Works, Box 7, Alfreton, Derbys.	17

Cementation Mining Ltd., Box 22, Bentley House, Doncaster DN5 OBT.	17, 47, 58, 62
Century Oils Ltd., Box 2, Hanley, Stoke on Trent, ST1 5HU.	38
Chapman Ltd., Clarke International Combustion Div., Sinfin Lane, Derby DE2 9GJ.	36, 39, 52, 55
Chrysler Internat., 68 Knightsbridge, London SW1	4,
Communication & Control Eng. Co. Ltd. Park Road, Claverton, Nottingham NG14 6LL.	13, 19, 42, 55
Compair Construction & Mining Ltd. Holman Works, Camborne, Cornwall TR14 8DS.	6, 15, 24, 25, 29, 40, 52, 56
Davis & Son Ltd., John, Box 38, Alfreton Rd., Derby, England.	13, 19, 53, 55
Davy Powergas International Ltd., 8 Baker St., London W1M 1DA.	16, 17, 31
Dosco Overseas Eng. Ltd., Ollerton Rd., Tuxford, Newark, Notts.	11, 18, 29, 54, 59
Dowty Meco Ltd., Meco Works, Bromyard Rd., Worcester WR2 5EG.	62
Dowty Mining Equipment Ltd., Ashchurch, Tewkesbury, Glos. GL20 8JR.	50, 54
Drill Sure Ltd., Froebel House, 21 Church St., Warwick.	34
Dust Suppression Ltd., Bourne End Mill, Hemel Hempstead, Herts HP1 2RW.	27, 29, 55
Eaton Corp, Waddensbrook Lane, Wednesfield, Wolverhampton WV11 3SW.	10, 45
Eimco (Great Britain) Ltd., Earlsway, Team Valley, Gateshead NE11 OSB.	24, 41, 42, 45, 56, 58
English Drilling Equipment Co. Ltd. Lindley Moor Rd., Huddersfield, Yorks HD3 3RW.	23, 24, 52
Euro-Drill Equipment (UK) Ltd., Derby Rd., Clay Cross, Chesterfield, Derbys.	23, 24, 25, 40
Fennel & Co. Ltd., J.H., Marfleet, Hull, Yorks HU9 5RA.	3, 20, 43
Fletcher Sufcliffe Wild Ltd., Universal Works, Horbury, Wakefield, Yorks WF4 5HR.	12, 20, 43, 50, 52, 54
GEC Mechanical Handling Ltd., Birch Walk, Erith, Kent DA8 1QH.	1, 6, 12, 16, 20, 21, 22, 26, 39, 42, 43, 52, 57, 63.

Golder, Hoek & Associates Ltd., 5/7 Forlease Rd., Maidenhead, Berks S16 1RP.	17,
Grantham Electrical Engineering Ltd. Harlaxton Rd., Grantham NG31 7SF.	20
Greenbat Ltd., Armley Rd, Leeds LS12 2TP	45
Greengate Industrial Polymers Ltd., Irwell Works, Ordsall Lane, Salford, Lancs.	3, 20
Gullick Dobson Ltd., Box 12, Ince, Wigan, Lancs WN1 3DD.	6, 41, 43, 50, 52, 54
Halifax Tool Co. Ltd., West Lane, Southowram, Halifax, Yorks HX3 9TW	24
Hayden Nilos Cornflow Ltd., Triumph Rd., Lenton, Nottingham NG7 2GF.	19, 29, 54, 55
Head Wrightson & Co. Ltd., The Friage, Yarm on Tees, Cleveland TS15 9DA	12, 16, 26, 39, 49, 64
Helipeds Ltd., Premier Works, Sisson Rd., Gloucester GL2 ORE.	39
Hudson (Raletrux) Ltd., Robert, Box 4, Morley, Leeds, Yorks LS27 8TG.	8
Hunting Surveys & Consultants Ltd. Elstree Way, Borehamwood, Herts WD6 1SB.	17, 31, 34
Huwood Ltd. Gateshead on Tyne, Tyne & Wear, NE11 OLP.	12, 20, 50
Hydraulic Drilling Equipment Ltd., Imperial Bldgs., Victoria Rd., Horley, Surrey.	23, 24
Jenkins of Retford Ltd., Member Babcock & Wilcox Ltd., Retford, Notts.	12, 16, 20, 22, 42
Johnson-Progress Ltd., Leek New Rd., Cobridge, Stoke-on-Trent, Staffs.	1, 3, 12, 20, 36, 52, 60
Joy Manufacturing Co. (UK) Ltd. Denver Equipment Div., Capitol House, 2 Church St., Epsom, Surrey.	1, 12, 16, 36, 37, 39, 52, 60
Lintott Engineering Ltd., Foundry Lane, Horsham, Surrey.	53, 55
Locker Industries Ltd., Box 161, Warrington, Lancs.	2, 20, 26, 31
Magco Ltd., Lake Works, Porcester, Fareham, Hamps PO16 9DS.	20, 21

Mastabar Mining Equipment, Church Bank Works, Church Accrington, Lancs.	52
Matbro Ltd. Matbro House, Horley, Surrey RH6 9EH	45
Matthew Hall, Ortech Ltd., 101-108 Tottenham Court Rd., London W1A 1BT	12, 16, 17, 64
Mining Dev. (UK) Ltd., Crown Lane, Howich, Bolton BL6 7NY.	1, 24, 41, 45, 52
Mining Supplies Ltd., Carr Hill, Doncaster, Yks.	6, 11, 12, 20
Mitchell Ropeways Ltd., Carr Hill, Doncaster, Yorks DN4 8DG.	17, 20, 48
Muir-Hill Ltd., Bristol Rd., Gloucester GL1 5RX.	45
Newell Dunford Eng. Ltd., Portsmouth Rd., Surbitton, Surrey KT6 5QF.	20, 21, 26, 27, 29, 39, 42, 51
Oldham & Son Ltd., Denton, Manchester M34 3AT.	55
Olin Energy Systems Ltd., North Hylton Rd., Sunderland, Durham SR5 3JD.	15, 39, 43, 52, 55, 62
Padley & Venables, Callywhite Lane, Dronfield, Sheffield S18 6XT.	6
Parsons Chain & Co. Stourport-on-Severn, Worcs. DY13 9AT.	42,
PD-NCB Consultants Ltd., 101-145 Gt. Cambridge Rd., Enfield Middx EN1 1UQ.	17, 47
Pegson Ltd., Coalville, Leics. LE6 3ES.	20, 21, 39, 52
Pickrose & Co. Ltd., Member Marmon Group Inc., Delta Works, Audenshaw, Manchester M34 5HS.	42
P & S Textiles Ltd., Div. Scapa Group Ltd. Broadway Mills, Haslingden, Lancs. BB4 4EJ.	27, 35
Ransomes & Rapier Ltd., Box 1, Waterside Works, Ipswich IP2 8HL.	32
RTZ Services Ltd., 6 St. James Sq., London SW17 4LD.	16
Ruston-Bucyrus Ltd., Excavator Works, Lincoln LN6 7DJ.	32
Scandura Ltd., Box 18, Cleckheaton, Yorks BD19 3UJ.	3

Schwartz-Holywell Ltd., Backworth, Newcastle-upon-Tyne, NE27 OAE.	52, 55
Sheepbridge Equipment Ltd., Chesterfield, Derbys S41 9QD.	8, 21
Simon-Warman Ltd., Todmorden, Lancs. OL14 5RT.	2, 19, 22, 39, 52
Soil Mechanics Ltd., Foundation House, Eastern Rd., Bracknell, Berks RG12 2UZ.	17, 34
Spencer & Sons Ltd., Matthias, Box 7, Arley St., Sheffield S2 4QQ.	24,
TBA Industrial Products Ltd., Belting Div., Wigan, WN2 4XQ	3
Thymark, Thyssen (Great Britain) Group, 8 The Sanctuary, Westminster, London SW1P 3JU.	16, 20, 41, 42
Torque Tension Ltd., Claylands Ave, Gateford Rd., Worksop, Notts S81 7BQ.	24
UMM Ltd., Box 19, Aycliffe Ind. Est. Darlington, Co. Durham.	6, 8, 10, 11, 12, 20, 21, 23, 24, 29, 46, 48
Underground Mining Machinery Ltd., Aycliffe Industrial Estate, Darlington, Durham.	8, 11, 20, 23, 42, 46
Westinghouse Brake & Signal Co. Ltd., Signal & Mining Div., Chippenham, Wilts SN15 1JD.	17, 19, 42
Wilfley Mining Machinery Co. Ltd., The Cambridge St., Wellingborough, Northants.	16
Wilkinson Process Linatex Rubber Co. Ltd., Stanhope Rd., Camberley, Surrey GU15 3BX	12, 16, 22, 39, 52, 64
Woodrow International Ltd., Taylor, Western Ave, London W5 1EU.	34
Zed Instruments Ltd., 76 Crown Rd., Twickenham, Middx TW1 3ET.	17, 19, 53

Fed. Rep. Germany

Ahlmann Maschinenenbau GmbH Postf. 725, D-2370 Rendsburg.	7
Auergesellschaft Sub., MSA Europe, Friedrich-Krause-Ufer 24, D-1000 Berlin 65	2, 31, 55
Aumund-Fordererbau GmbH, Saalhoffer Str. D-4134 Rheinberg.	3, 20, 28, 32, 45
Bavaria Maschinenfabrik GmbH & Co. Industriestr. 34, D-7910, Neu-Ulm.	57, 63

Becorit Grubenausbau GmbH, Postf. 209, D-4350 Recklinghausen	41, 43, 50 54
Bochumer Eisenhutte Heintzmann & Co. Bessemerstr. 80, D-4630 Bochum.	50, 54
Brown Boveri & Cie AG, D-6800 Mannheim.	3, 13, 17, 42
Buckau-Wolf Maschinenfabrik AG, Postf. 10 04 60, D-4048 Grevenbroichl.	5, 12, 17, 20, 21, 31, 47
Buhler-Miag, Ernst-Amme-Str. 19 D-3300 Braunschweig.	20, 21, 22, 26, 39
CEAG-Dominit AG, Munsterstr. 231, D-4600 Dortmund.	44
Clouth Gumminwerke AG, Postf. 60 02 29, D-5000 Cologne 60.	3, 17, 20
Continental Gummi-Werke AG, Postf. 169, D-3000 Hanover 1.	3, 20, 48
Deilmann-Haniel GmbH, Postf. 130220 D-4600 Dortmund 13	42, 45, 58
Demag AG, ABT Bergwerksmachinen, Wolfgang-Reuter-Platz, D-4100 Duisburg	15, 17, 18, 20, 24, 32, 34, 40, 42, 42, 58
Demag Baumaschinen GmbH, Postf. 180180, D-4000 Dusseldorf 13.	32
Demag Fordertechnik GmbH, D-5802 Welter-Ruhr.	10, 42
Demag Drucklufttechnik GmbH, POB 900360, D-6000 Frankfurt 90.	15, 24
Demag Lauchhammer Masch & Stahlbau GmbH Postf. 230, Forststr., D-4000 Dusseldorf 1.	3, 5, 20, 32, 45
Demag Verdichtertechnik GmbH, Postfach 10 01 41, D-4100 Duisburg 1.	15, 17, 18, 20, 24, 32, 34, 40, 42, 58
Deutsche Montabert GmbH, Nauroder Str. 23, D-6200 Wiesbaden.	6, 24
Dragerwerk AG, Moislinger Allee 53-55, D-2400 Lubeck	28, 55
Eastman International Co. GmbH, Carl-Zeiss Str. 16, D-3005 Hanover	33
Eickhoff Maschinf bk-U Eisengiesserei Mb, Postf. 629, D-4630 Bochum.	11, 19, 20, 45, 46, 59
Esch-Werke AG, Postf. 100807, D-4100 Duisburg 1.	20, 21, 26, 39, 63

Faun-Werke Nurnberg, Postf. 8, D-8560 Lauf AD Pegnitz.	55
Flottmann-Werke GmbH, Postf. 1240/1260, D-4690 Herne 1.	15, 24, 40
Friemann & Wolf GmbH, Meidericherstr. 6-8, D-4100 Duisberg.	44
Frieseke-Hoepfner GmbH, D-8250 Erlangen-Bruck.	33, 52
Gutehoffnungshutte Sterkrade AG, Postf. 110240, D-4200 Oberhausen 11	8, 15, 17, 20, 42, 45, 47, 56, 58, 62
Hauhinco Maschinenfabrik Zweigerstr. 28-30, Box 639, D-4300 Essen, 1.	40, 43, 52
Hazemag GmbH & Co., Dr. E. Andreas, Rosnerstr. 6-8 Postf. 3447, D-4400 Munster	20, 21, 26, 31, 36, 50
Hoechst AG, Postf. 800320, D-6230 Frankfurt 80	38
IBAG International Baumaschinenfabrik AG Box 308, Branchweilerofstr. 35, D-6730 Neustadt.	3, 12, 17, 19, 20, 21, 36, 39, 63
KHD Industrianlagen AG, Humboldt Wedag. Postf. 910404, Wiersbergstr., D-5000 Koeln 91	1, 4, 6, 9, 12, 16, 20, 21, 22, 26, 27, 37, 39, 49, 51, 52, 57, 60, 62, 63
Klemm, Gunter, Consulting Engineer D-5962 Drolshagen.	43
Knorr-Bremse GmbH, Moosacher Str. 80 D-8000 Munich 40	15
Krauss-Maffei AG, Krauss-Maffei Str. 2, D-8000 Munich 50	9, 26, 36
Krupp GmbH, Fried, Krupp Industrie-Und Stahlbau Postf. 141960, D-4100 Duisburg 14.	1, 3, 9, 16, 17, 21, 26, 27, 32, 34, 36, 37, 39, 47, 51, 57, 60, 64
Kulenkampff Gebruder, Postf. 103869, D-2800 Bremen.	20, 35, 39
Langen & Sondermann, Postf. 1628, D-4670 Lunen.	50
Liebherr Hydraulikbagger GmbH D-7951 Kirchdorf.	7, 32
Loesche KG, Postf. 5226, Steinstr. 18, D-4000 Dusseldorf 1.	39
Lurgi Gesellschaften, Gervinusstr. 17-19, D-6000 Frankfurt/Main.	16, 22, 27, 37, 49
M.A.N., Katzwanger Str. 101, D-8500 Nurnberg	17, 20, 32, 52

Montan Consulting GmbH, Postf. 5, D-4300 Essen.	17, 47
Montan-Forschung Hofstr. 56-60, D-4010 Hilden.	13, 34, 55
Motoren-Werke Mannheim AG, Postf. 1563, D-6800 Mannheim 1.	55
Nilos GmbH, Forderband-Ausrustung, Achenbachstr. 26, D-4000 Dusseldorf.	3, 20
O & K Orenstein & Koppel AG Post. 170167, Karlfunkestr., D-4600 Dortmund.	3, 6, 21, 32, 42, 45, 52, 57
O & K Orenstein & Koppel AG, Werk Ennigerloh, Postf. 25, D-4722 Ennigerloh.	20, 21, 26, 39
Pleuger Unterwasserpumpen GmbH, Friedrich-Ebert-Damm 105, D-2 Hamburg 70.	52
Pohlig-Heckel-Bleichert Ver. Maschin. AG., Pohligstr. 1, D-5000 Koeln 51.	5, 10, 20, 42
Polysius Werke, Postf. 340, D-4732 Neubeckum.	5, 17, 22, 26, 34, 39, 57
Prakla-Seismos GmbH, Postf. 4767, Haarstr., D-3000 Hanover 1.	17, 23, 33, 34
Rheinbraun-Consulting GmbH, Stuttenweg 1, D-5000 Koln 41.	17, 34, 47
Ritz Pumpenfabrik KG, Postf. 188 D-7070 Schwaebisch.	52
Rost & Co., H. Balatroswerke, Postf. 901168, D-2100 Hamburg 90.	3, 20
Rud-Kettenfabrik Rieger & Dietz, Postf. 1650, D-708 Aalen.	20
Saarberg-Interplan, Postf. 73, Stengelstr. 1, D-6600 Saarbrucken.	17, 34, 47
Salzgitter Maschinen AG, Postf. 511640, D-3320 Salzgitter 51.	12, 41, 42, 45, 56
Scholtz AG, Conrad, Am Stadstrand 55-59, D-2000 Hamburg 70.	3
Schopf Maschinenbau GmbH, Postf. 93, D-7000 Stuttgart 75.	8, 56
Schuberth-Werk AG, Postf. 5029, D-3300 Braunschweig.	55

Schwartz & Dyckerhoff AG, Ruhrthaler Maschinfabrik, Box 1260, Scheffelstr. 14-28, D-4330 Mulheim-Ruhr.	63
Siemag Transplan GmbH, Postf. 1270/1280, D-5902 Netphen 1.	42, 58
Siemens AG, Postf. 3240, D-8520 Erlangen	13, 17, 19, 42, 52
Soding & Halbach, J.C. Edelstahlwerke D-5800 Hagen.	23
Standard Filterbau GmbH, Roesnerstrasse 6/8, D-4400 Munster.	12, 17, 20, 22, 27, 28, 29, 35, 62, 64
Thiele, August, Kettenfabrik, Postf. 8040, D-5860 Iserlohn-Karthof.	20
Thysen Industrie AG, Umform Bergbautechnik, Erlinger Str. 80, D-4100 Duisberg 28.	11, 50, 54
Thysen Schachtbau GmbH, Postf. 011480, D-4330 Mulheim-Ruhr.	17, 58
Titanit Bergbau Technik, Postf. 1347, D-4702 Heessen.	11,
Ulm GmbH Maschinenbau, Box 1249, D-7910 Neu Ulm.	14
Wagener & Co., Box 218, inder Graslake 20, D-5830 Schwelm.	3
Wasag Chemie Sythen GmbH, Rolundstr. 9, D-4300 Essen 1.	4
Westfalia Lunen, Postfach, D-4670 Lunen.	6, 11, 20, 21, 41, 45, 46, 54
Wirth & Co. KG Alfred, Postf. 1327, D-5140 Erkelenz 1.	23, 24, 25, 52
Zeiss Carl, Postf. 1369, D-7080 Oberkochen	34

Finland

Ahlstrom Oy, A, Varkaus Engineering Works, SF-78100 Varkaus.	26
Exel Oy, Uunisepantie 7, SF-00620, Helsinki 62.	55
Outokumpu Oy, Technical Export Div. Box 27, SF-02201 Espoo 20.	2, 17, 37, 39, 47
Rauma-Repola Oy, Lokomo Div., Box 306-307, SF-33101 Tampere 10.	20, 21, 32
Rauma-Repola Oy, Pori Div., Pori.	1, 7, 21, 22, 26, 37, 39, 60, 62

Roxon Oy, SF-15860 Salpakangas	3, 6, 16, 17, 20, 21, 22, 31, 36 37, 55, 57, 60, 64
Tampella AB, Oy, Tamrock Div., SF-3310 Tampere 31.	15, 24, 29, 45, 56

France

Alsthom Atlantique, Div. Neyrpic, 75 rue General Mangin, F-38100 Grenoble.	3, 12, 16, 17, 21, 22, 26, 31, 39, 52, 60, 63, 64
APC Azote & Produits Chimiques SA, 62-68 rue Jeanne D/Arc, F-7546 Paris 13.	4
APOD, 17 rue Docteur Grenier, F-25300, Pontarlier.	48
Berry Co., 92 rue Bontepollet, Lillie 59000.	12, 16, 26, 48, 62
BK, Cie, Francaise, 15 rue de Billancourt, F-92100 Boulogne S/Seine.	21
Boyer SA, BP 28, F-02105, St. Quentin.	12, 20
Compagnie Electro Mecanique, Sub. Brown, Boveri & Co., 12 rue Portalis, F-75383 Paris 8.	3, 13, 17, 19, 42
Dragon SA, Appareils, Box 11, F-38600 Fontaine.	20, 21, 39, 63
Fives-Cail Babcock, 7 rue Montalivet, F-75383 Paris 8.	1, 5, 6, 9, 12, 15, 16, 17, 19, 20, 21, 22, 26, 39, 42, 52, 57, 58, 60, 64
Foraco, 24 Ave George V, F-75008 Paris.	23, 33
France Loader, 50 Ave. Victor Hugo, F-75116 Paris.	8, 45, 48, 56
Joy SA, 209 Rue de Bercy, F-75585 Paris 12.	8, 21, 24, 42, 56
Maco-Meudon, Chemin de Genas, F-69800 St., Priest.	6, 15, 23, 24, 27, 40, 42, 52, 58
Minex Mine-Expert, 35 Champs-Elysees, F-75008 Paris.	8, 11, 17, 43, 46, 47, 50, 52, 54, 59
Normet SA, 83 rue Montmartre, F-75002 Paris.	8, 48
Oldham France SA, BP 99, 62 Arras	2, 13, 55
OTP Engineering, 35 rue Volta, F-92801 Puteaux.	17

Poclain, F-60330 La Plessis, Belleville.	7, 14, 32, 43, 52.
Realization Equipments Industriels, 14 rue D'Annam, F-75020 Paris.	5, 12, 20, 32, 45, 56.
S.A.M.I.I.A., BP 127, 2 Place Maugin, F-59506, Douai.	10, 42, 55
Saulas & Cie, 16 rue Du Buisson St. Louis, F-75010 Paris.	12, 35, 57, 60
Secmafer S.A., BP 42, F-78203 Mantes	34, 43
Secoma, 274 Cours Emile Zola, F-69100 Villeurbanne.	8, 24
SILEC, Signalisation Industrielle Div., 69 rue Ampere, F-75017, Paris.	13, 19, 42, 55
Sofreimines, 59 rue de la Republique F-93108 Montreuil Sous Bois.	17, 34, 47
Stein Industrie 19 Av. Morane-Saulnier Velizy-Villa coublay.	26, 36, 39, 57
Stephanoise de Constr. Mecaniques, Soc., BP 519, F-42007 Ste-Etienne.	6, 11, 17, 20, 21, 59

Hungary

Aluterv, Pozsonyi ut. 56, POB 128, H-1389 Budapest.	17, 20, 21, 36, 52, 60
Banyaterv Mining Design Institute, POB 83, H-1525 Budapest.	17, 47
Budavox Telecommunication Co. Ltd. POB 267, H-1392 Budapest	13, 17, 19
Chemolimpex, Hungarian Trading Co., Box 121, V Deak Ferqnc U 7-9, H-1805 Budapest	3, 20
Geominco, POB 92, H-1525 Budapest.	17, 34, 58, 64
Nikex, Joszef Nador Ter. 5-6, H-1809 Budapest 5.	1, 3, 21, 22, 26, 37, 39, 42, 47, 50, 60, 64

India

McNally Bharat Engineering Co. Ltd. Kumardhubi 828203, Dist. Dhanbad.	6, 12, 16, 17, 20, 21, 22, 26, 37, 39, 57, 63

Iran

South African Iranian Mining Co., Box 12, 1366 Tehran.	17,

Italy

ARMCO Moly-Co. SpA CP116 1-33043 Cividale Del Friuli.	39
Del Monego SpA, Piazza Della Repubblica 8, 1-20121 Milan.	22. 26
Reiter & Crippa SNC, Via Roverto 3, 1-20059 Vimercate.	7, 12, 17, 20, 21, 26, 36

Japan

Bando Chemical Industries Ltd., International Div., 2 Nagaoka Bldg., 8-5, 2-Chome, Hatchobori, Chuo-Ku, Tokyo 104.	3
Furukawa Rock Drill Sales Co. Ltd., 6-1 Marunouchi 2-Chome, Chiyoda-Ku, Tokyo.	6, 19, 22, 24, 27, 40, 45, 52, 55, 58
Hitachi Construction Machine Co. Ltd., 2-10, 1-Chome, Uchikanda, Chiyoda-Ku, Tokyo 101.	32, 45
Hitachi Ltd., Nippon Bldg., 6-2, 2-Chome, Ohtemachi, Chiyoda-Ku, Tokyo 100.	10, 13, 15, 19, 23, 52, 53
Ishikawajima-Harima Heavy Industries, 2-1, 2-Chome, Shin-Ohtemachi, Ohtemachi, Chiyoda-Ku, Tokyo.	1, 4, 5, 6, 7, 9, 12, 15, 16, 17, 19, 20, 21, 26, 30, 31, 32, 39, 42, 49, 50, 51, 54, 55, 57.
Kawasaki Heavy Industries Ltd., 2-4-1, Hamatsu-Cho, Minato-Ku, Tokyo.	15, 17, 22, 26, 27, 32, 36, 39, 45, 49, 52, 55, 57, 60
Kawasaki Heavy Industries Ltd., Machine Sales Div., 4-1 Hamamatsu-Cho, 2-Chome, Minato-Ku, Tokyo.	1, 7, 15, 21, 22, 23, 26, 27, 31, 32, 37, 39, 45, 52, 55, 57, 60, 62, 64.
Kobe Steel Ltd., Tekko Bldg. Tokyo.	21, 26, 27, 32, 39, 45, 57.
Koken Boring Machine Co., 20-13, 2-Chome Taira-Machi, Meguro-Ku, Tokyo.	17, 23, 24, 25, 34, 52.
Komatsu Ltd., Akasaka 3-6, 2 Chome, Minato-Ku, Tokyo 107.	8, 14, 15, 45
Kurimoto Iron Works Ltd., 2-11, 2-Chome, Nihonbashi, Chuo-Ku, Tokyo.	21, 22, 26, 27, 39, 57, 63
Mitsubishi Metal Corp., World Trade Center, Bldg., Hamatsu-Cho, Minato-Ku, Tokyo.	17, 22, 47, 52
Mitsubishi Steel Mfg. Co. Ltd., 9-31 Shinonome 1-Chome, Kohton-Ku, Tokyo.	7, 24, 36, 39
Mitsuboshi Belting Ltd., 4-7 Hamazoedori, Nagataku, Kobe.	3, 20, 45

Mitsui Shipbuilding & Eng. Co. Ltd., 4-14 Tsukiji 5-Chome, Cho-Ku, Tokyo.	24
Sanyo Chemical Industries Ltd., 11-1 1KKO Nomoto-Cho, Higashiyama-Ku, Kyoto 605.	38
Seiki Kogyosho Ltd. (SKK), 1-Chome Higashi-Tsukaguchicho, Amagasaki 661.	42
Shinko Electric Co. Ltd., 12-2-3 Chome, Nihonbashi, Chuo-Ku, Tokyo 103.	19, 20, 45
Tokai Rubber Industries Ltd., 1-1-12 Akasaka, Minato-Ku, Tokyo.	3
Toyo Kogyo Co. Ltd., Fuchu-Cho,AkiGun, Hiroshima	24
Yokohama Rubber Co. Ltd., 36-11 Shimbashi, 5-Chome, Minato-Ku, Tokyo 105.	3

Mexico

Gates Ruber de Mexico SA de CV, Ave. de las Torres, 226, Naucalpan de Juarez Mexico, DF, Mexico.	3, 20, 45

Netherlands

Conrad-Stork, Box 134, Haarlem.	23, 24

Northern Ireland

Davidson & Co. Ltd., Sirocco Engineering Works, Belfast BT5 4AG.	62,

Norway

Den Norske Remfabrik A/S Box 1, N-1410 Kolbotn	3, 20, 39, 55
Kvaerner Brug A/S, Kvaernerveien 10, Oslo 1.	39
Noratom-Norcontrol AS, Holmenvn 20, Oslo 3.	19
Thune-Eureka AS, Box 225, Oslo 1.	36

Peru

Arce Helberg, Jose E. Petit Thouars 4380, Miraflores, Lima 18.	34

Philippines

ARMCO Marsteel Alloy Corp., Box 1528, Commercial Center, D-708 Makati, Rizal.	39

Poland

Kopex, Export-Import, Grabowal, 40-952 Katowice.	2, 4, 6, 11, 12, 13, 16, 17, 18, 19, 20, 21, 24, 27, 28, 29, 36, 37, 42, 46, 47, 50, 52, 55, 58, 62.

Roumania

Geomin Calea Victoriei 109, Bucharest 22.	1, 3, 17, 20, 21, 23, 24, 25, 26, 33, 34, 39, 47, 52

RSA

Barlows Heavy Engineering Ltd., Box 183, Benoni, Tvl. 1500.	24, 26, 39, 42, 58, 60
Bateman Ltd., Edward L., Box 565, Boksburg, Tvl.	1, 6, 9, 11, 12, 16, 17, 18, 19, 20, 21, 22, 26, 27, 36, 37, 39, 46, 47, 49, 50, 52.
Ore Sorters Africa Pty. Ltd., Box 781138, Sandton, Tvl. 2146.	16, 57
Osborn (South Africa) Ltd., Samuel, Box 25619, Denver, Tvl.	6, 7, 12, 16, 21, 22, 39, 51
Scaw Metals Ltd., Box 61721, Marshalltown Tvl. 2107.	7, 39, 52
Vecor Heavy Engineering Ltd., Box 9442, Johannesburg, 200 Tvl.	8, 21, 26, 32, 39, 42, 58

Scotland

Alluvial Dredges Ltd. Blackhall Lane, Paisley.	16,
Anderson Mavor Ltd., Div. Anderson Strathclyde Ltd., Box 9, Flemington Works, Motherwell, Lanarks ML1 1SN.	11, 13, 18, 19, 20, 32, 45, 46, 59, 61
Consolidated Pneumatic Tool Co. Ltd., Fraserburg, Aberdeens. AB4 5TE.	14, 15, 23, 24, 52
Flexible Ducting Ltd., Cloberfield, Milngavie, Glasgow G62 7LW.	62
Howden & Co. Ltd., James 195 Scotland St., Glasgow G5 8PJ.	62
Nobel Explosives Co. Ltd. Nobel House, Stevenston, Ayres KA20 3LN.	4,

Spain

Taim S.A., Calle Aldebaran, Zaragoza 12.	17, 20

Sweden

Asea Inc., Mining & Industrial Transport Avd., FMK, S-72183 Vasteras.	19, 42, 55
Atlas Copco AB, S-10523 Stockholm.	1, 6, 15, 23, 24, 28, 33, 42, 45, 52, 56
Atlas Copco ABEM AB, Box 20086, S-16120 Bromma.	33
Bofors, AB, Steel Div., S-69020 Bofors.	7
Hagglund & Soner AB, Trucks & Mining Machinery Div., Fack, S-89101 Ornskoldsvik.	8, 20, 45
Kockum Industri AB, Fack, S-26120 Landskrona.	61
Linden-Alimark AB, Fack, S-93103 Skelleftea.	24, 42, 58
LKAB International AB, Fack, S-10041 Stockholm.	17, 34, 47
Morgardshammer AB, Fack, S-77701 Smedjebacken	21, 39, 52
Persona Sparteknik AB, Box 101, S-27100 Vstad.	8
Pneumatisk Transport AB, Box 32058, S-12611 Stockholm.	4
Pumpex AB, Skebokvarnsvagen 370, S-12434 Bandhagen.	52
Sala International, Box 137, S-73300 Sala.	1, 16, 20, 22, 26, 36, 37, 39, 52,
Sala Minco, Box 137, S-73300 Sala.	1, 12, 16, 20, 22, 36, 37, 42, 52, 55, 57, 60, 64.
Skega AB, S-93040 Ersmark	39
Stenberg-Flygt AB, Fack, S-17120 Solna 1.	52
Svedala-Arbra AB, S-23300 Sevedala	20, 21, 39
Transtronic AB, Postf. 175, S-73101 Koping.	17, 19, 55

Trelleborgs Gummifabriks AB, Fack, S-23101 Trelleborg 1.	3, 17, 20, 31, 39, 45, 52
Volvo BM AB, S-63185	32, 45, 56
Wennberg AB, C.J. Fack, S-65101 Karlstad 1.	17

Switzerland

Brown Boveri & Co. Ltd. Postf. 84, CH-5401 Baden.	3, 13, 17, 19, 42
Meyhall Chemical AG, CH-8280 Kreuzlingen.	38
SIG Schweizerische Industrie Gesellshaft, CH-8212 Neuhausen Am Rheinfall	24, 40, 58

U.S.A.

Abex Corp., 530 5th Ave., New York, NY 10036.	7, 19, 39, 43, 52
Acker Drill Co. Box 830, Scranton, PA 18501.	23, 24, 52
AEC Inc., Box 106, State College, PA 16801.	27
Agar Instrumentation Inc. 2320 Blalock, Houston, TX 77080.	2, 31, 55, 62
Air Products & Chemicals Inc. Box 538, Allentown, PA 18105.	5, 51, 55, 57
Allis-Chalmers, Crushing & Screening Div. Box 2219, Appleton, WI 54911.	21
Allis-Chalmers, Cement & Mining Systems Box 512, Milwaukee, WI 53201.	15, 19, 20, 21, 22, 29, 39, 45, 52
American Air Filter Co. Inc. Box 1100, Louisville, KY 40201.	11, 22, 27, 29, 36, 49, 57, 62
American Biltrite Rubber Co. Inc. Boston Industrial Products, Cambridge MA 02139.	3, 20
American Cyanamid Co., Mining Chemicals Dept., 859 Berdan Ave, Wayne, NJ 07470.	30
American Mine Door, Box 6028, Sta. B. Canton,OH 44710.	55, 62

American Optical Corp., International Div. 14, Mechanic St, Southbridge, MA 01550. — 55

American Precision Industries, Dustex Div. Box 900, Greeneville, TN 37743. — 27

AMI Corp. 212 Washington Ave,, Box 5475, Towson, MD 21204. — 43. 52

Armco International, Div. Armco Steel Corp. Box 700, Middletown, OH 45042. — 39

Ashland Chemical Co., Chemical Products Div., Box 2219, Columbus, OH 43216. — 38

Athey Products Corp, Box 669, Raleigh, NC 27602. — 12, 20, 45

Auto-Dustrial Inc. 310 Northern Blvd., Great Neck, NY 11021. — 55

Auto-Weigh Inc., Box 4017, 1439 N. Emerald Ave., Modesto, CA 95352. — 20

Autometrics, 4946 N. 63rd St., Boulder, CO 80301. — 2, 19

Babcock & Wilcox Co., Automated Machine Div. 1 Northfield Plaza, Troy, ML 48084. — 19

Baldwin Mfg. Co., J.A., Kearney, NB 68847. — 36

Ballagh & Thrall Inc., 1201 Chestnut Street, Philadelphia, PA 19107. — 3

Barber-Greene Overseas, Overseas Div., 400 N. Highland Ave, Aurora, IL 60507. — 20, 32

Barrett, Haentjens & Co. Hazleton, PA 18201. — 52

Becker Drills, Inc. 5055 E. 39th Ave. Denver,CO 80207. — 23, 34

Beckman Instruments Inc., 2500 Harbor Blvd. Fullerton, CA 92634. — 2, 19, 31, 33, 55

Bemis Co., Inc. Box 188, 800 Northstar Center, Minneapolis, MN 55402. — 27

Bethelehem Steel Corp., Bethlehem, PA 18016. — 8, 54

Betton Inc., 2846 N. Prospect St., Colorado Springs, CO 80907. — 26, 36, 52

Bico Inc. Box 6339, Burbank, CA 91510 — 21, 39, 51

BIF Co., Div. General Signal Co. Box 217, West Warwick, RI 02893. — 19, 36, 52

Bin-Dictator Co., 1915 Dove, Port Huron, MI 48060.	19
Bird Machine Co. Inc., Neposet St., S. Walpolke, MA 02071.	9, 31
Brooks Mfg., Inc., Frank, Box 7, Bellingham, WA 98225.	60
Brown & Root Inc., Box 3, Houston, TX 77001.	17, 47
Bucyrus-Erie Co., Box 56, South Milwaukee WI 53172.	24, 31, 42
Calcon Corp., Box 1346, Calgon Center, Pittsburgh, PA 15230.	35, 38, 64
Calweld Div. Smith International Inc. 9200 Sorenson Ave, Box 2875, Sante Fe Springs, CA 90670.	8, 20, 25, 42
Card Corp. Box 117, Denver CO 80201.	6, 7, 42, 48, 58
Carpco, 4120 Haines St., Jacksonville, FL 32206.	16, 20, 57
Case Co., J.I. International Div., 700 State St., Racine, WI 53404.	14, 31, 45
Caterpillar Tractor Co., 100 NE Adams St., Peoria, IL 61629	7, 14, 31, 45
CE Raymond/Bartlett-Snow, Div. Combustion Engineering, 200 W. Monroe St., Chicago, IL 60606.	22, 26, 51, 57
Centrifugal & Mechanical Ind. Inc., 146 President St., St. Louis, MO 63118.	9
CF & I Steel Corp., Box 1830, Pueblo, CO 81002.	39
Challenge-Cook Bros. Inc., 15421 E. Gale Ave., City of Industry, CA 91745.	41
Chapman, Wood & Griswold Inc., 4015 NE Carlisle Ste, E. Albuquerque, NM 87107.	12
Chicago Pneumatic Tool Co., Drill Div. Box 1225, Enid, OK 73701.	24
Cincinnati Mine Machinery Co., 2980 Spring Grove Ave., Cincinnati, OH 45225.	20
Clark Equipment Co., Construction Machinery Div., Box 547, Benton Harbor, MI 49022.	7, 17, 32, 45

Clark Equipment Co., Fargo Div., 112N University, Fargo, ND 58102.	45
Clark Equipment Co., Lima Div., Lima, OH 45802.	32
Clark International Marketing SA, Box 333, Benton Harbor, MI 49022.	6, 7, 32, 45
Cleveland Wire Cloth & Mfg. Co. The 3573 E 78th St., Cleveland, OH 44105	35
Columbia Steel Casting Co., Inc. Box 03095, Portland, OR 97203.	39
Connellsville Corp, Box 677, Connellsville, PA 15425.	20, 42
Coors Porcelain Co., 600 9th St., Golden, CO 80401.	21, 39
Dart Truck Co., Box 321, Kansas City,MO 64141.	7, 45
DCE Vokes Inc., 10101 Linn Station Rd Ste. 900, Jeffersontown, KY 40223.	27, 29, 30, 31, 36
Deere & Co., John Deer Rd, Moline, IL 61265.	7, 32
Deister Concentrator Co., Inc. 925 Glasgow Ave, Fort Wayne, IN 46801.	12, 16
Deister Machine Co. Inc., Box 5188, 1933 E. Wayne St., Fort Wayne, IN 46805.	20
Dillon & Co. Inc., WC, Box 3008, 14620 Keswick St., Van Nuys, CA 91407.	19, 42, 55
Dings Co., Magnetic Group, 4761 W. Electric Ave, Milwaukee, WI 53219	16
Dixon Valve & Coupling Co. 1201 Chestnut St. Philadelphia, PA 19102.	15
Donaldson Co. Inc., Box 1299, Minneapolis, MN 55440.	27, 28, 29, 36
Dorr-Oliver Inc., 77 Havemeyer Lane, Stamford, CT 06904.	1, 9, 12, 16, 22, 26, 36, 39, 52, 57, 60, 64.
Dow Chemical Co., The, Functional Prods. & Sys. Dept., 2020 Abbot Rd., Midland MI 48640.	4, 52
Dravo Corp, 1 Oliver Plaza, Pittsburgh, PA 15222.	17, 42
Dresser Industries Inc., Industrial Equipment Div., 601 Jefferson, 28th Floor, Houston, Texas 77002.	3, 6, 10, 12, 20, 21, 26, 36, 51

Dresser Industries Inc., Industrial Products Div., 900 W. Mount St., Connersville, IN 47331.	10, 15, 52, 62
Dresser Industries Inc. Mining Services & Equip. Div, Box 24647, Dallas, TX 75224.	23, 24, 25, 27, 52
Drilco Industrial Div, Smith International Inc., Drawer 3135, Midland, TX 79702.	23, 24, 25, 61
Driltech Inc., Box 338, Alachua, FL 32615.	23, 24
Du Pont de Nemours & Co., Inc. E.l., Wilmington, DE 19898.	2, 4, 9, 17, 30, 31, 33, 55
Eaton Corp., 100 Erieview Plaza, Cleveland, OH 44114.	3, 10, 19, 42, 45, 56
Eimco Mining Mach., Div. Envirotech Corp. 669 W 2nd South, Salt Lake City, UT 84110.	16, 35, 36, 60, 64.
Elmac Corp. of New Mexico, Box 1056, 506E Center St., Carlsbad, NM 88220.	8, 20, 48, 56
Engineered Equip. Co., Box 2395, Santa Ana, CA.	7
Eriez Magnetics Ltd., 395 Magnet Dr. Erie, PA 16512	20, 49, 57
Esco Corp., 2141 NW 25th Ave., Portland, OR 97210.	7, 12, 21
Fate-Root-Heath Co., The Plymouth Locomotive Works, Bell St., Plymouth, OH 44865.	45
Federal-Mogul Corp., International Distribution Div., Box 1966, Detroit, MI 48235.	36
Fiat-Allis Corp, Box 1213,Milwaukee, WI 53201.	7, 32, 45
Fisher Scientific Co. 711 Forbes Ave., Pittsburgh, PA 15219.	2, 55
Flexible Steel Lacing Co., 2525 Wisconsin Ave., Downers Grove, IL 60515.	3
Flexowall Corp., One Heritage Park, Clinton, CT 06413.	3, 20
FMC Corp., Construction Equipment Div., 1201 SW 6th St., Cedar Rapids, IA 52406.	20, 32
FMC Corp., Material Handling Equip. Div. 898, Lexington Ave., Homer City, PA 15748.	20
Foxboro Co., Neponset Ave., Foxboro, MA 02035.	19

Frontier Kemper Constructors, Box 6548, Evansville, IN 47712.	42, 58
Fuller Co., Box 29, Catasauqua, PA 18032	15, 20, 21, 22, 26, 27, 39, 52, 57, 62
Galigher Co., The Box 209, Salt Lake City, UT 84110.	1, 12, 37, 39, 52
Gardner Denver International, Box 4714, Dallas, TX 75247	15, 23, 24, 40, 42, 52
Gates Rubber International, 999 S Broadway, Denver, CO 80217.	3
General Electric Co., Instrument Products Div., 40 Federal St., West Lynn, MA 01910.	19, 53, 55
General Mills, 4620 W. 77th St., Minneapolis, MN 55435.	38
General Motors, Terex Div., Hudson, OH 42236.	45
General Splice, Box 158, Rte 129, Croton-on-Hudson, NY 10520.	3
Geometrics, 395 Java Dr., Sunnyvale, CA 94086.	33, 34
Geophysical Service Inc., Box 5621, MS 970, Dallas, TX 75222.	34
Getman Corp., Box 549, South Haven, MI 49090	48
Golder Associates, Inc., 10628 NE 38th Place, Kirkland, WA 98033.	17, 34, 47
Goodall Rubber Co. 430 Whitehead Rd, Trenton, NJ 08604.	3, 20
Goodman Equipment Corp., 4834 S. Halstead St., Chicago, IL 60609.	18, 20, 61
Goodrich International, BF, 500 S. Main St., Akron, OH 44318.	3, 20, 35, 39, 55
Goodyear International Corp, 1144 E. Market St., Akron, OH 44316.	3
Gould & Co. Gordon I, 235 Montgomery St., Ste 2300, San Francisco, CA 94104.	16, 17, 26, 31, 34, 47
Gruendler Crusher & Pulverizer Co. 17 N. Market St., Ste. 2915, St. Louis, MO 63106.	1, 6, 20, 21, 22, 51, 57, 63
Gundlach Machine Co. TJ, 1 Freedom Dr., Belleville, IL 62222.	21

Hamilton Engineering Associates, 16765 SW 32nd Ave., Seattle, WA 98166.	17
Hammermills Inc., Sub. Pettibone Corp. 626 NW C. Ave. Cedar Rapids, IA 52405.	20, 21, 51
Hanson Co. RA, Mining Div., Box 7400, Spokane, WA 99207.	20
Harnischfeger Corp., Box 554, Milwaukee, WI 53201.	19, 42
Hazen Research International Inc., 4601 Indiana St., Golden, CO 80401.	1, 16, 17, 20, 26, 34, 36,
Heath & Sherwood Co. Box 468, 3800 W 5th Ave, Hibbing, MN 55746.	16, 39, 40, 51
Heil Process Equipment Corp. 34250 Mills Rd. Avon, OH 44011.	52, 57, 62, 64
Heinrichs Geoexploration Co. Box 5964, Tucson AZ 85703.	17, 31, 33, 34.
Hendrick Mfg. Co. 70 Dundaff St. Carbondale, PA 18407.	35
Hensley Industries Inc. 2108 Joe Field Rd., Dallas, TX 75220.	7
Hewitt-Robins, Box 1481, Columbia, SC 29202.	3, 21
Heyl & Patterson Inc., 7 Parkway Center, Pittsburgh, PA 15220	6, 9, 12, 16, 17, 20, 22, 26, 37
Holz Rubber Co., 1129 S. Sacremento St., Lodi, CA 95240.	3
Honeywell Inc. Process Control Div., 1100 Virginia Dr., Fort Washington, PA 19034	19, 53,
Hossfield Manufacturing Co. 448 W. 3rd St., Winona, MN 55987.	23, 24
Hughes Tool Inc., Box 2539, Houston, TX 7701	58
Humphreys Engineering Co. 818 17th St. Denver, CO 80202.	16
Huwood Irwin Co., Box 409, Irwin, PA 15642.	6, 8, 11, 17, 19, 20, 46, 50, 52, 54, 59
Ingersoll Rand, Co., Woodcliff Lake, NJ 07675	6, 15, 23, 24, 25, 40, 42, 52, 58
International Engineering Co. Inc., 220 Montgomery St. San Francisco, CA 94104.	17

International Harvester Co., 401 N. Michigan Ave. Chicago, IL 60611.	32, 45
Iowa Mold Tooling Co. 500 Hwy 18 West, Garner, IA 50438	15
Ireco Chemicals Co. 726 Kennecott Bldg. Salt Lake City, UT 84133	4
ITT Marlow, Box 200, Midland Park, NJ 07432.	4, 19, 52
Jacobs Engineering Co. 837 S. Fair Oaks Ave. Pasadena, CA 91105	17, 31
Jaeger Machine Co. The, 550 W. Spring St., Columbus OH 43216.	15, 24, 52
Jeffrey Mining Machine Co. Div. Dresser Industries Inc. 274 E 1st Ave, Columbus OH 43216.	11, 13, 18, 20, 29, 45, 46, 50, 52, 54, 56, 59, 62
Jordan Associates, Box 14005, Oklahoma City, OK 73114.	15, 23, 24
Joy Mfg. Co., Oliver Building, Pittsburgh, PA 15222.	1, 2, 8, 11, 12, 15, 16, 18, 21, 23, 24, 27, 37, 39, 40, 42, 45, 46, 48, 49, 52, 56, 59, 62.
Joy Manufacturing Co. Denver Equipment Div. 7503 Marin Dr., Box 22598, Denver, CO 80222	1, 12, 16, 17, 26, 36, 37, 39, 52, 60, 63
Joy Mfg. Co., Western Precipitation Div. Box 2744 TA, Los Angeles, CA 90051.	19, 22, 28, 49, 62
Kaiser Engineers, 300 Lakeside Dr., Oakland CA 94666.	17, 31, 34, 47, 64
Kay-Ray Inc., 516 W. Campus Dr. Arlington Heights, IL 60004.	2, 19
Kennedy Van Saun Corp., Sub McNally Pittsburg Mfg. Co. Beaver St., Danville, PA 17821.	12, 20, 21, 26, 39, 52
Kent Air Tool, 711 Lake St., Kent, OH 44240	6
Koehring Co. 780 N. Water St., Milwaukee, WI 53201.	14, 23, 32, 44
Koppers Co. Inc. Hardinge Oper., Box 312, York, PA 17405	21, 26, 36, 39
Krebs Engineers, 1205 Chrysler Dr., Menlo Park, CA 94025	12, 22
Lawrence Pumps Inc., 371 Market St., Lawrence MA 01843	52

Lebus International Engineers, Inc. Box 2352, Longview, TX 75601.	42
Leco Corp. 3000 Lakeview Ave. St. Joseph, MI 49805.	2
Lee Norse Co. Charleroi, PA 15022	18, 59
Leeds & Northrup Co. Summeytown Pike, North Wales, PA 19454.	2, 19, 53
Lifts Unlimited, 2498 Newbury Dr. Cleveland OH 44118	10
Lincoln St. Louis, Div. McNeil Corp., 4010, Goodfellow Blvd., St. Louis, MO 63120.	15, 36, 52
Logan Actuator Co. 4956 N. Elston Ave., Chicago, IL 60630.	53, 55
Long-Airdox Co., Div. Marmon Group Inc. Box 331, Oak Hill, WV 25901	3, 4, 6, 12, 15, 17, 18, 20, 24, 48
Longyear Co. 925 S.E. Delaware St., Minneapolis, MN 55414	23, 24, 34
Lug-All Co., 538 Lancaster Ave., Haverford, PA 19041	42
Marathon Letourneau Co. Longview Div., Box 2307, Longview, TX 75601.	45
Marathon Manufacturing Co. 600 Jefferson, Houston, TX 77002.	7, 56
Marion Power Shovel Co. Inc. 617 W. Center St. Marion, OH 43302.	7, 24, 32
Martin-Decker Co. 1928 S. Grand Ave. Santa Ana, CA 92705.	17, 19, 33, 53
Martin Engineering Co. Rte 34, Neponset IL 61345	4
Materials Control Inc. 719 Morton Ave., Aurora, IL 60506	19
McDowell-Wellman Engineering Co. 113 NE St. Clair Ave., Cleveland, OH 44114.	1, 5, 17, 31, 45
McKee & Co. Arthur G. Western Knapp Engineering Div. 2855 Campus Dr. San Mateo, CA 94403.	17
McNally Pittsburg Mf. Corp. Drawer D. Pittsburg, KS 66762.	12, 16, 20, 21, 26, 52

Merrick Scale Mfg. Co. 180 Autumn St., Passaic, NJ 07055.	36
Mine Safety Appliances Co. 600 Penn Center Blvd., Pittsburgh, PA 15235	2, 13, 28, 36, 41, 55, 62
Mine & Smelter Co. Sub Barber-Greene Co. Box 16067, 3800 Race St., Denver, CO 80216	12, 16, 21, 39, 51, 63
Mintec International Div. Barber-Greene Co. 400 N. Highland Ave., Aurora, IL 60507	17, 20, 21, 32
Mobile Drilling, 3807 Madison Ave. Indianapolis, IN 46227.	23, 24
Mountain State Engineers, Box 17960, Tucson, AZ 85731.	12, 16, 17, 31
Mueller Co. 1201 Chestnut St., Philadelphia, PA 19107.	55
Nagle Pumps, Inc. 1250 Center St., Chicago Heights, IL 60411.	52
National Filter Media Corp. 1717 Dixwell Ave. Hamden, CT 06514.	3, 35
National Iron Co. Div. Pettibone Corp. W 50th Ave & Ramsey, Duluth, MN 55807.	20
Nat Mine Service Co. 3000 Koppers Bldg. Pittsburgh, PA 15219.	8, 18, 44
National Tank & Pipe Co. Box 17158, Portland OR 97217	60
Niro Atomizer Inc. 9165 Rumsey Rd., Columbia, MD 21045	26
Norsk Hydro Sales Corp. 800 Third Ave., New York, NY 10022.	4
Northwest Engineering Co., 201 W. Walnut St. Green Bay, WI 54305.	32
Ohmart Corp. The, 4241 Allendorf Dr. Cincinnati OH 45209.	19
Olko Engineering, 500 Fifth Ave. New York NY 10036	17
Over Lowe Co. 2767 S. Tejon St., Englewood CO 80110.	14, 44, 62
Owatonna Tool Co. 809 Eisenhower Tower, Owatonna, MN 55060.	52
Owen Bucket Co., 6001 Breakwater Ave., Cleveland, OH 44102.	7

Paccar International Inc. Box 1518 Bellevue, WA 98009	45
Page Engineering Co. Clearing Post Office Chicago, IL 60638	7, 32
Parsons Co. The Ralph M, 100 W. Walnut, Pasadena, CA 91124	17
Parsons-Jurden Co. Div. Ralph M. Parsons Co. 100 W. Walnut St., Pasadena CA 91124.	17, 34
Peabody International Corp., ABC Group Box 187, 330 Kings Hwy., Warsaw, IN 46580	62
Peake Cecil V. 365 Oxford Dr. Bethlehem, PA 18017.	17
Pecoff Bros., Nursery & Seed Inc. Rte 5, Box 215, R. Escondido, CA 92025.	55
Peerless Pump, 1200 Sycamore St., Montebello, CA 90640.	52
Pekor Iron Works Inc., Box 909, Columbus, GA 31902.	22
Pennsylvania Crusher Corp., Box 100, Broomhall, PA 19008	6, 21
Permutit Co. The Sub, Sybron Gorp. 49E Midland Ave., Paramus, NJ 07652.	36, 64
Pettibone Corp., 4700 W. Division St. Chicago, IL 60651.	14, 32, 45, 52
Philadelphia Gear Corp., 181 S. Gulph Rd. King of Prussia, PA 19406.	1
Phillipi Hagenbuch, Inc., 1815 N. Knoxville Ave, Peoria, IL 61603.	17,
Phoenix Products Co. 4745 N. 27th St. Milwaukee, WI 53209.	44
Pincock Allen & Holt Inc. 4420 E. Speedway Blvd., Tucson, AZ 85712.	17
Portadrill, 2201 Blake St., Denver, CO 80205.	23, 24
Portec Inc. Pioneer Div. 3200 S.E. Como Ave., Minneapolis, MN 55414.	20, 21, 36, 63
Programmed & Remote Systems Corp., 899 Hwy 96W, St. Paul, MN 55112.	39
Prosser Industries, Div. Purex Corp, Box 3818, Anaheim, CA 92803.	52

Raygo Inc. 9401 N. 85th Ave, Minneapolis MN 55445	14
Redpath Ltd., J.S. Box 1176, Casa Grande, AZ 85222.	17, 58
Reed Tool Co. Drilling Equipment Div. Box 90750, Houston, TX 77090.	24
Reinco Inc. Box 584, Plainfield, NJ 07061	30
Reliance Electric Co., Dodge Div. 500 S Union St., Mishawaka, IN 46544.	3
Rexnord Inc. Process Machinery Div., Box 383, Milwaukee, WI 53201.	20, 21, 39, 42
Rexnord Inc. Vibrating Equipment Div. Box 13007, 3400 Fern Valley Rd, Louisville, KY 40213.	5, 12, 20
Riblet Tramway Co. Box 5220, Spokane, WA 99205.	48
Robbins Co. The 650 S. Orcas St. Seattle WA 98108.	24, 25
Roberts & Schaefer Co. 120 S. Riverside Plaza Chicago, IL 60606.	12, 22
Round & Son Inc, David, 32405 Aurora Rd. Cleveland, OH 44139.	10, 42
Rupp Co., The Warren, Box 1568, Mansfield, OH 44901.	52
Salem Tool Co. The 767 S. Ellsworth Ave, Salem, OH 4460.	24
Saverman Bros. Inc. 620 S. 28th Ave., Bellwood, IL 60104.	32, 42
Schramm Inc. 901 E. Virginia Ave. West Chester, PA 19380	15, 23, 24, 40
Seegmiller Associates, 447 E. 200 South, Salt Lake City, UT 84111	17, 47,
Serpentix Conveyor Corp. 1550 S. Pearl St. Denver, CO 80203	20
Simplicity Engineering Co. 212 S. Oak St. Durand, MI 48429.	20
Smith International Inc., 4667 Macarthur, Newport Beach, CA 92660.	23, 24, 25
Smith Tool Co. Div. Smith International Inc. Box C-19511, Irvine, CA 92713.	24

Southern Iowa Mfg Co. Drilling Products Div. Box 448, Osceola, IA 50213.	23, 24, 43
Sprague & Henwood Intl. Corp. 221 W. Olive St. Scranton PA 18501.	34
Square D, Co. 205 S. Northwest Hwy, Park Ridge, IL 60068.	19, 44
S & S Seeds Inc. 382 Arboleda Rd, Santa Barbara, CA 93110	30
SSP Construction Equipment, Miller-Viber Products Box 2038, Pomona, CA 91766	14
Stanadyne Inc., Harford Div., Box 1440, Hartford, CT 06102.	36
Stanco Manufacturing & Sales, 800 Spruce Lane Dr., Harbor City, CA 90710.	19, 45, 52
Stansteel Corp., Sub. Allis-Chalmers Corp., 5001 Boyle Ave., Los Angeles, CA 90058	22, 26
Stearns Magnetics Inc. 6001 S. General Ave. Cudahy, WI 53110.	16
Stearns-Roger Inc. Box 5888, Denver CO 80217.	17, 26, 31, 47
Stephens-Adamson Mfg. Co. Ridgeway Ave., Aurora, IL 60507.	20
Straub Mfg. Co., Inc. 8383 Baldwin St., Oakland, CA 94621.	21, 39
Sturtevant Mill Co., 20 Sturtevant St. Boston, MA 02122.	21, 51, 57
Superior Fiber Products Co. 501 Executive Plaza, Hunt Valley, MD 21031.	30
Telsmith, Div. Barber-Greene Co., 532 E. Capitol Dr. Milwaukee, WI 53201.	21, 22, 63
Terradex Corp,, 1900 Olympic Blvd. Walnut Creek, CA 94596	34
Teton Exploration Drilling Co. Inc. Sub United Nuclear Inc., 254 N. Center, Casper, WY 82601.	34
Texas Instruments Inc. Box 5621, MS 956, Dallas, TX 75222.	34
Texas Nuclear Corp. Div. Ramsey Engineering Co. Box 9267, Austin, TX 78766.	2
TRW Mission Mfg. Co. Box 40402, Houston, TX 77040.	24, 52

Twin Disc Inc. 1328 Racine St., Racine, WI 53403.	19
Uniroyal International Inc. Oxford Center, Middlebury, CT 06749.	3, 12
Unit Rig & Equipment Co. Box 3107 Tulsa, OK 74101	45
Universal Oil Products, Johnson Div. Box 3118, St. Paul, MN 55165.	12
Valsan International Corp. 43W 61st St. New York, NY 10023.	52
Varian Associates, Instrument Div. 611 Hansen Way, Palo Alto, CA 94303.	2
Vibra Screw Inc., 755 Union Blvd., Totowa, NJ 07512.	5, 20, 36, 64
Vibranetics Inc. 2714 Crittenden Dr. Louisville, KY 40209	20, 26
Viking Explosives & Supply Inc. Star Rte 2 Box 16A, Hibbing, MN 55746.	4, 52
Wagner Mining Equipment Co. Box 20307, Portland, OR 97220	56
Wahler & Associates, WA Box 10023 Palo Alto, CA 94303	17
Weir Co., Paul, 20N Wacker Dr. Chicago, IL 60606.	17, 47
Wemco, Div. Envirotech Corp, Box 15619, Sacramento, CA 95813	1, 9, 12, 16, 22, 37, 52,57
Western Gear Corp. Heavy Machinery Div. 2100 Norton Ave., Everett WA 98201.	42, 52
Wheelabrator-Frye, Inc., Air Pollution Control Div., 600 Grant St., Pittsburgh, PA 15219.	27, 31, 35, 36, 49, 64
Wilfley & Sons, Inc., AR, Box 2330, Denver, CO 80201.	52
Wilson Products Div, Div. ESB Inc. 2nd & Washington St., Reading, PA 19603	52, 55
Wilson LK (Ken) Box 7123, Menlo Park, CA 94025	34
Winchester-Western, 275 Winchester Ave, New Haven, CT 06504.	26
Worthington Compressors Inc. 333 Elm St., West Springfield, MA 01089	24

Worthington, Box 1250, Mountainside, NJ 07092.	15, 52
Zeni Drilling Co. 324 Eighth St. Morganton, WV 26505.	17, 23, 25
Zip-Up Inc. Box 3147 Crs. Rock Hill, SC 29730.	44

USSR

Machinoexport V/O, Mosfilmovskayo 35, Moscow V-330	7, 16, 18, 20, 21, 22, 24, 32, 33, 34, 37, 39, 45.
Techmashexport V/O, Mosfilmovskaya Ul, 35 Moscow V-330.	36, 52, 62

Wales

Angus & Co. Ltd., George, Belting Products Div., Sloper Rd., Cardiff, CF1 8TD.	3, 20
Barker, Davies & Co. Old Bank Chambers, Pontypridd, Glam.	42, 58
Brown, Lenox & Co. Ltd., Pontypridd Glam CF37 4BY.	21
CPMS Co. Ltd. Thyssen (Great Britain)Group Bynea, Llanelli, Dyfed.	4
Flexadux Plastics Ltd. Thyssen (Great Britain) Group, Bynea, Llanelli, Dyfed.	29, 62
Reliance Rope Attachment Co. Ltd. 27 Park Place, Cardiff CF1 3QL	42, 58
Robertson Research International Ltd. Tyn-Y-Coed, Llanrhos, Llandudno, Gwynedd LL30 1SA.	17, 34, 47
Thyssen (Great Britain) Ltd. Bynea Llanelli, Dyfed, SA14 9SU.	4, 12, 17, 29, 34, 35, 41, 42 47, 58, 62.

West Nigeria

Metal & Minerals (Nigeria) Ltd. Box 222, A 198 IsoKun, Ilesha.	17

BIBLIOGRAPHY

The following papers were of considerable help in providing an expert background to the study, in addition to three Coal publications, Coal Age, Colliery Guardian and World Coal. Those manufacturers which provided literature describing their range of equipment are also listed after the general bibliography. A considerable amount of information was provided too, by the national information centres listed in Appendix G, to which should be added Departments of State reponsible for energy matters.

Assessment of future vehicle developments for underground use based upon the concept of the self propelled train - Giro Mining Transport

Application of computer techniques to coal preparation - A.H. Parker & S. Robertson U.K.

Aromatic Chemicals from Coal - K.O. Schowalker and E.F. Petras

Automation in Coal Mining - J. Olaf F.G.R.

Automation device for face alignment - M. Noel, France

Automatic laser plumbs - F.R. Tolman, National Physical Laboratory U.K.

L'Avenir du Charbon - Ministere de l'Industrie et de la Recharche 1976

British Gas Corporation Annual Reports

Chemical Feedstocks from Coal - C.D. Frohning & B. Cornhils Ruhrehemie AG

CO_2 Acceptor Progress - George Curran, Conoco Coal

Coal Liquefaction in U.K. - J.S. Harrison, N.C.B.

Coal Preparation - H.M. Spanton

Comparison and Co-ordination of national policies and programmes in energy R & D Sector - F.R.G.

Computer-Controlled Communications and Monitoring - H. Dobroski & H.E. Parkinson, Pittsburgh

Co-ordination of policy for science and technology - Ireland - National Science Council Report

Current techniques in deep shaft sinking and development - H.A. Longden, U.K.

Dept. of Energy (U.K.) National Conference proceedings 1976

Drilling - a new concept by K. Shaw

Drilling underground roadways - H.F. Watson

Energy prospects - Cavendish Laboratory Cambridge 1976

Employee involvement in health and safety measures in coal mining - J. Gormley, National Union of Mineworkers

Energy in Western Europe - CEPCO

Energy R & D Programme of the Government of the Federal German Republic Annual Reports - 1974, 1975-1976

European Coal - CEPCO

Eurostat - Coal

European Communities Publications and Statistics

First revision of energy policy programme for Federal Republic of Germany
Fluidised Bed Gasification of various coals - S.J. Gasior, A.J. Forney, W.P. Haynes & R.F. Kenny, Pittsburgh
Fluidised Combustion - Combustion Systems Ltd. U.K.
Fluidised Combustion of Fossil Fuels - BCURA, U.K.
Future dependence on coal - Leslie Grainger, U.K.

H-Coal Pilot Plant Programme - H.H. Stotler, U.S.
Health in coal mines - Dr. C. Amoudru, France
Horizon control systems designs for longwall facing machines - Wolfenden & Hartley

Improvement in coalmine gas detection instrumentation - G.H. Schnakenberg & M.D. Aldridge, U.S.A.
International Energency Agency - Leslie Grainer

Le Charbon, energie nouvelle de demain
Le gasification souterraine du Charbon - P Ledent, INIEX
Legislative, Institutional & Financial Factors - J.B. Brennan, NCB Washington, DC.

Mechanisation, Monitoring & Control of Processes - R.B. Dunn, U.K.
Medium - Btu Gas, its importance and meaning - M.C. Goodman 1977

NCB Publications
- Coal
- Coal for the future
- Coal's new face
- Coal in U.K.
- Coal Research Centre Reports 1976 & 1977
- Island of Coal
- Medical Service Reports 1976-7
- Mining beyond 2000 A.D.
- Mining Research & Establishment Report 1977
- New promise for underground gasification
- Opencast coalmining
- Production & Productivity bulletins
- Prospects for expansion in coal
- Report and Accounts 1974/5, 1975/6, 1976/7
- Your project is Coal

Objectives & priorities for research & development projects on remote control & monitoring - W. Brand, EEC
OECD statistics 1978
Official Reports in Parliamentary Debates - U.K.
Oxygenated Hydrocarbon from coal based synthesis gas - Peter H. Spitz, New York 1976

Proceedings of 20th Congress of Mine Safety & Health Commission, Luxembourg
Proceedings of 13th World Gas Conference 1976
Progress in Hygas - Frank C. Schora, IGT

Rack-a-track - J.M. Mills - NCB
Rapport D'Orientation sur la R & R en Matiere d'Energie
Recent powered support developments - J.J. Graham, Dowty Mining Equipment
Remote control in mining machinery - A. le Fevre

Sensors & Instrumentation for control and monitoring in coal mines - L.R. Cooper, U.K.
Solvent Refined Coal Process - Baldwin, Golden & Garry, U.S.A.

Trade & Industry - U.K.
Transportation - J.H. Northerd, U.K.
Typical colliery in the year 2000 - P.G. Tregelles, MRDE

U-Gas Process - J.W. Loading & J.G. Patel, IGT
Underground gasification of coal - P.N. Thompson, J.R. Mann & F. Williams, NCB
Underground gasification - Medaets, Brussels
UNO World Energy supplies 1975
United States energy scene 1977 - J.F. O'Leary
U.S. Bureau of Mines approach to automation in coal mining - W.B. Schmidt, M.L. Lavin & W.R. Griffiths, U.S.A.

Uses of Water in the winning and transport of minerals - Heinz Harnisch

Westfield Development - B. Hebden, British Gas Corporation
World Energy Conference Reports

<u>Manufacturers Literature</u>

Alkerströms Bjorbo AB
Anderson Strathclyde
Auriema
Berry - France
British Jeffrey Diamond
Cementation Mining
Compair
Davis of Derby
Dowty Mining Equipment
Gullick Dobson
Gyro Mining Transport
Hunslett
Hydraulic Drilling Equipment
ICI Nobles Explosives
J.C.B.
Jeffrey Dresser
Jenkins of Retford
Mining Supplies Ltd.
Paxman
Piteraft
Siemag
Westfalia Lünen

INDEX

This index has been compiled basically under subject heads. Chemical Production - Coal - Explosives - Faults - Gasification - Health & Safety - In-situ - Liquefaction - Logging - Mining methods. Similar references are to be found under national heads.

Otherwise individual references are to be found for Manufacturers, Organisation Processers and items of special interest or those not falling automatically under the major heads.